HOW TO USE AURA SCANNER

IMAGINATION IS NOW A REALITY, UNVEIL THE HIDDEN

GAURAV KUNAL

Made with ❤ on the Notion Press Platform
www.notionpress.com

My prayers to the scientists and the Saptrishis.

Dedicating this research book to my workplace - **Bikham Healthcare, Mr. Harman Dhawan & Sonia Dhawanmam**. I am blessed that I got the opportunity to work with such beautiful people here at Bikham.

Mr. & Mrs. Dhawan

Contents

Contents

Contents

Foreword

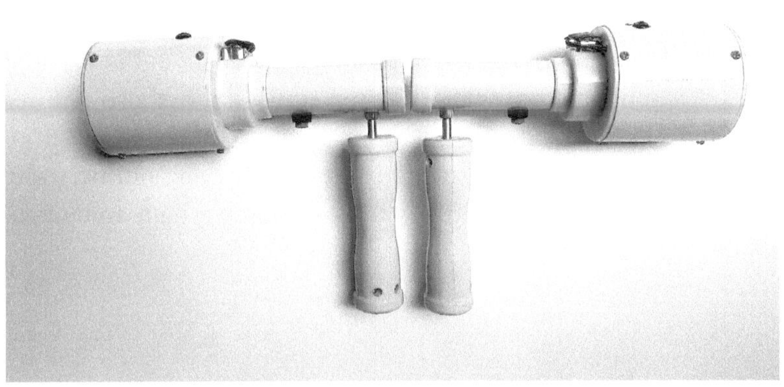

Metal Body Aura Scanner

Disclaimer: Aura Scanner will only give correct answers once you get proper training in it. After the training, it's all about practice. Also if you learned how to operate it but you leave it idle for 2-3 months and then suddenly one day you start taking the reading, in that case, it won't work well. The training is provided through online paid workshops. Also, I would like to mention at the very beginning itself that different manufacturers' device working criteria will be different in the same way as every mobile's function is to make calls but then a Samsung phone will be different from a Redmi phone. I recommend the one from Manglam Vastu, Haridwar due to its accuracy.

There is a very big reason to write this book. The unfortunate thing in India is that people easily give way to blind faith. Let a so-called astrologer tell you that you have a Vastu Dosha in your home and from that day you will build ways for that person to bring forward more and more negative points and frighten you. This belief is baseless. If you have a Vastu Dosh then it must be from a particular cause and when you remove the cause it should show the result that everything is fine now. so you can only check that with an Aura Scanner. there is no other method to see what is the positivity of your house at a certain phase of time.

I have seen people giving wrong readings to people in the name of Aura Scanning, not only their method is illogical but also reduces the image of Astrologers. Be wise to decide whether you want to follow someone's teaching or not. Just don't let any ordinary guy come from nowhere and fool you by charging a high amount of money for remedies.

About The Author

Gaurav Kunal was eager towards Astrology and the Occult since his college times. He is a BSc. Graduate in Jewelry Designing. Gemology was his favorite subject and so became Astrology. Currently posted in Bikham Healthcare as Medical Biller, he practices general Aura Reading consulting at his leisure. He is the second of the two sons of the former transport businessman **Mr. Sanjay Singh & Mrs. Indira Singh.**

Kunal was initiated into Astrology as an amateur hobby of Palmistry. Later he found that astrology and Palmistry are vague science. You can find a lot of disputes between astrologers reading the same birth chart. It is not the case in Aura Scanning. The latest device which Kunal uses is manufactured by Mr. Yogesh Keshwani from Mangalam Vastu, Haridwar. The aura reading machine manufactured by Mangalam Vastu is 100% accurate and is time-tested. There are various workshops and one-to-one pieces of training one of which was attended by Kunal himself as his initial training. After this, he also took part in two more workshops to know this science deeper. Rest is all about practice and dedication.

There are three types of aura scanners - Fiber body Aura Scanner, Metal body Aura Scanner, and Brass body Aura Scanner, among which I chose the metal body Aura Scanner, which is also 100% accurate. One needs to purchase the scanner after which his practice practically begins. The more he practices the more he excels in prophecy. No one is a master from the mother's womb. The only option is to learn through practical experiences.

During his experiences, he found many people were interested in getting correct predictions and not vague predictions as is usually a practice among horoscope readers. There was a flood of people approaching online as well as through persons from various parts of the country. The common problem among them was the lack of correct analysis of their Janam - Kundli. They reported that they were harassed, misguided, and better call looted by the so-called Astrologers who claimed themselves as Masters but failed in even basic Faladesh. To earn money by building fear in Jataka is sadly now a common practice in India. Some malefic Yogas which are partially present in the Kundli are presented in a very rigorous way which inculcates fear among the people and they are forced to pay to ward off the Evil. This is a very sad but harsh reality.

Here is when Aura Scanner comes to the rescue. You don't need to be an astrologer to get these readings. The scanner gives you an answer in a simple Yes / No format. If the aura scanner predicts that the gemstone is suitable for you then it is. If the aura scanner predicts you are going to a foreign land next year then it will surely happen. If the Aura scanner gives the result that a person is having Pitra Dosha then definitely he has that. Aura Scanner will help you overcome it by performing one of the solutions you ask for. In my understanding, there is by far no more accurate technique of prediction than using an Aura scanner.

In the initial stages, taking no fee, working day and night, and entertaining every single query was a strenuous task. People are not willing to give room to a new path. But luckily this was not the case with this section of astrology for Kunal. He got recognition very soon among his known and unknowns, and not less than 50 queries in a day became a usual happening. You can imagine the curiosity for this new device among the mass. The best thing about out is that it shows the result to the client as well as the Aura Reader himself. **In astrology, the horoscope reader concludes that he conveys to the common person in a language of Stars which generally no one understands. No one knows what is meant by Gaj Kesari Yog, BudhAaditya Yog or Viprit Rajyog, etc. However, the results of Aura Scanner are visible to both parties. There is complete transparency and understanding between the client and the Astrologer. This is a very positive point for this Aura Reading science.**

Among the various hobbies of the author - music, painting, and astrology, Aura Reading holds a completely different value. Even at night 3 o'clock, he might be found reading the Aura of someone's photograph at his place. There is no Good time or Bad time for Aura reading. Kunal has scanned from fruits to paintings, from Visiting cards to flute voice and even plants. Kunal believes in research activities. Aura Scanning is a great Renaissance in astrology, whether it is believed by mass today or the day after tomorrow. It is a unique science.

So be it.

Acknowledgements

Thanks to my parents for supporting me as this path is untrodden in my family. Thanks to **Mr. Saurav Kunal**, my elder brother, mentor, and supporter, who silently gives a push to all my strange endeavors.

My sincere thanks to **Manglam Vastu (Haridwar)**, as that is the company, from where I got the training to operate the Aura Scanner and got this magical device. **Mr. Yogesh Keshwani**, Manglam Vastu (Haridwar) was the person who inspired me toward this science.

There is a video on youtube in which he has demonstrated **how we can know from an Aura scanner that a person is Dead or Alive**, from any unknown person's photograph. That video inspired me to learn this science. Yogesh sir often says, "**If you learn how to operate this machine, you can get the answer to anything**". I would like to say from my side that there are many options for a person to purchase the machine but among them, the most accurate Aura Scanner and best guidance that we can rely upon is Manglam Vastu. **Mr. Yogesh Keshwani**, the Managing Director, is an honest, well-behaved person and I have no second opinion about his predictions.

Aura reading is not child's play but also it is not a very tough thing. After the training, the person will be able to read all the hidden and visible aspects of the Aura of any person's photo printout. Plenty of people approach aura scanning and after they get satisfactory results they also give positive feedback. I am thankful to everyone who approached me and gave their valuable feedback.

People are eager to know more and more because there is no end to this science. When I read the aura of a person I keep in mind that I give a solution to anyone's problem through this machine and not just by guessing which many Tarot Card readers and Horoscope readers do.

On September 2022 a gentleman approached me. He was curious to know about this future also he wanted me to predict his present life. Now first of all we need to call the person in front of us before the scanning device or the second option is that we have to get the photo printout of the named person whom we want to check. When I got the photograph and I checked, the results were a bit disappointing. I could see the planetary positions for the positivity of the planets was average, he was married but when I checked for the question of children, I concluded that the person is not having a child. The aura scanner gave an affirmation on the marriage query denoting that the person was married but when I put the sample of childbirth it showed negative reading. The person was childless at that time. Also as I mentioned the professional life of the person was average I informed him as per my readings. As expected, he was shocked. Then he wanted to know about the remedy. Now the beautiful thing about the scanner is that it not only shows you the result of the future, the present but also, will tell you what if you do a certain act of remedy in your life in the future. For example, if I ask whether the worship of Lord Hanuman boon a child for him? **The answer will be either Yes or No. When we want to check about the possibility of an event the answer is that time will be in percentage!** This small incident doubled my belief in Aura Scan. Also, I'm thankful to him for his frankness.

There is a complete scarcity of books on Aura Scanning. Very few people know about aura scanning. Lesser people can do correct Aura Scanning/reading. If you search on YouTube you will get a lot of videos about this, which is misleading. I used the word misleading because astrologers nowadays are focused more on earning or better say looting the person rather than helping them through genuine reading. He should tell you what is going to happen to you and not just tell you to do a remedy so that bad things don't happen. The objective should be to give a prediction and not to provide a solution for an incident before the client asked for it. This Aura scanning machine which I have bought from a reliable manufacturer tells me the actual status of the Planets of a person. So, I will be helping the person by understanding the planetary positions. What other aura readers do is that they don't want to accept the reality that their machine is not of the latest technology and accurate as it should be. The reason might be the fault of the manufacturers. Also, the most important thing is that at least 4 Chakras should be balanced when you do the reading. If it is not balanced you will never get the correct answers.

I have seen the videos of Aura Scan in which a person performs Aura Scanning on a person's photo opened on a mobile, which is wrong because the input here is a mobile screen whose frequency changes rapidly which is why the reading will not be accurate. Some other astrologers perform reading on a video call, on the screen of a laptop. This is merely a publicity stunt it never gives accurate results. That is why the client will get no benefit from it. This is just to fool viewers and earn money from them. They will tell you all the Doshas and out of fear you will also think to get the remedies. If by chance in the future you get any failure in the ventures, you will think that The Astrologer was correct and if by chance you get success The Astrologer will tell you that it was because of the remedy which he suggested. That is why astrology is a field that is full of doubts.

A common person wants a solution but is not able to have faith in every astrologer. Every next Astrologer tells him a different story. No two astrologers will tell you the same points. They will either tell you contradictory things or tell you two different observations. Whom to trust? But yes, you can rely on Aura scanning because the reading is the same every time. Aura reading is not complex but it will need a little experience, training, and knowledge. You can become a successful Aura Reader if you follow the steps which I am going to mention in this book.

My sincere thanks to my apprentice who gave me true feedback. I will not say I was right in my readings all the time but yes the feedback which I will mention here is as true as The Sun in the sky, I mean to say that there is 0% editing in the feedback. The clients were so happy when I told them the answers to their questions and that's why decided I should manuscript this.

Names worth mentioning :

Murari Sinha Sir

Debopam Mukhopadhyay

Er. Deepak Swaraj

Rahul Dev

Sonu Bhaiyya

Ambresh Arya

Akshay Arya

Dr. Shubhranshu Tiwari

Thanks & praise to the Lord of the Lords, Mahakal.
Jay Shiv! Har Har Mahadev sang Parvati...!
Jai Baba Baidyanath!

Gaurav Kunal

c/o Sanjay Singh, H No : 103, Mandaar,
opp Bank of Baroda, Carstairs Town,
B. Deoghar, Jharkhand – 814112
Mob – 8757237101

Testimonials

"The Aura Scanning experience was awesome, brother you answered very well to all my questions and cleared all my doubts, especially the Gemstone scan was too good, I was confused about which one will suit me best and your readings helped me a lot. The readings were crystal clear and genuine. You are an angel, thank you bro _/_"

- Rrahul Surre

"You are doing great help to people by letting them know the truth with the help of Aura science rather than going to dhongi baba ji. Thank you Kunal for help yaara... Grateful to you for everything."

- Bhardwaj, K.

"Apki sari bate bilkul proof ke sath sahi thi. Jitna bhi aapne dikhaya sab puri tarah sahi hai. Aapke bataye hrah sthiti aur Mantra sabhi ki information sahi nikli vo bhi proof ke sath. Thankyou so much. Really happy and thankful for all that _/_ _/_"

- Taneja, D.

"The readings of Aura Scanning were done patiently and with accuracy. All the answers I had asked him were given to me. Even I had asked twice. I would recommend him for true readings as he has everything to check and check your questions accurately... I am blessed to get in touch with him. Thank you so much, Gaurav Ji"

- Yadav, P.

"Bohot sara tension kam hua aisa lag raha hai.... Muhe to lag raha tha ki main aise hi berojgar rahunga... Aapne positivity bhar di

_/_"
- Ashutosh Ji

"Nice! Aura Scanning! Recommend it for everyone! To burst misconceptions and get the right guidance to improve life! Nice attitude of Gaurav !"
- Dr. Bhardwaj, Rini

"Dear Kunal Astro, I'm really thankful for your Aura Scan reading. This Aura Scanner reading is very helpful for me... I am very happy... Thank you so much Kunal ji _/_"
- Sharma, R.

"All correct! Kudos. You are on the next level!!!"
- Mukhopadhyay, Debopam

Brass Body Aura Scanner

True Stories

Some of my most interesting Real Encounters with Aura Scanner

CHAPTER ONE

The Mango Leaf

One day I thought I should give this Aura machine a test. The day before, I performed some experiments with plants. The experiment was that whichever plant's leaf I put in the sample box of the scanner the Aura Scanner would start pointing toward that plant's direction! It might sound a bit normal but if you saw it in reality you would think how is it happening? This Aura Scanner identifies the Aura of the plant whose leaf is chosen and differentiates it from other plants. I just wanted to test it because in my home I know the results already but what if I don't know the answer? Will this Aura scanner pass this kind of test?

So without wasting time, I went to my best friend's house, which has got a lot of beautiful plants and almost thick greenery in his garden. I also didn't know how many and which all trees he had. But this aura Scanner has the magical quality that it will tell which plant's leaf we have taken as a sample. What I did is that I closed my eyes and asked my friend to take any plant's leaf and put it inside a sample box. I was clueless about which plant's leaf was chosen but this scanner had the potential to find it out. Someone was passing by and said hello and I failed the first attempt. I gave myself a second chance again he went somewhere and came back with different leaves in the box. I couldn't see the leaves also but I had to find out which plant he had selected. Slowly I lifted the scanner and moved around. It pointed to the front and then when I went to a mango leaf it stopped. When I brought the scanner on top of the mango plant it fully opened stating that the leaf or better say DNA was of the mango tree. My friend was highly impressed and said. "I don’t usually believe in these things but here I agree with your findings. It is a Mango leaf!"

CHAPTER TWO

Where to do Boring?

I have uploaded a separate video on this topic on my youtube channel. There is a story behind it. One night I was on break while on duty. I was working from home, 1 a.m. Suddenly I thought I should operate the scanner. It was drowsy. at that time I was involved in finding out the location of the tea in the room placed by me myself. Instead of tea, it came to my mind let's experiment with water. This idea came to me due to the rain. So I took an empty sample box and filled it with drinking water. When I entered the room my scanner was on my hand and when I raise it, it was pointing towards a bottle filled with drinking water, kept on the window! I checked this by standing at different positions from the water bottle. Where I went suppose if I stood at the right side of the bottle, the scanner would point to the left - towards the bottle. In the same way, if I stood at the left side of the bottle, it would point toward the bottle to the right of me. This was quite interesting. I took the bottle and kept it on the floor then took many readings as I wanted. And when I used to put the scanner on top of it then it would fully open. That was a lucky day for me. I thought I can not miss this so at 3:30 a.m just after finishing my office work I made a demo video for this. I could not wait till morning.

In the morning I came out to my garden area where there is a Well.

I wanted to check whether this could detect the well water. I had used Well water in the sample at that time. Yes! it was giving a clear indication of where is the well, from as far as 50m. So I concluded the video by showing the demo of well water too.

Boring should be done at a place where the aura scanner is fully open while using the sample of Boring from Sample box 2. Then after that, again reading is to be taken with the sample of tap water at the same spot. If this time also the scanner opens then it is confirmed that water will be found if someone does boring at that spot.

Experiment with Drinking Water

Aura Scanner fully opened on Well

CHAPTER THREE

Shaapit Neelam

As Aura videos started gaining popularity more people wanted to know what is their Energy status. Due to lack of time, I could perform only a few readings and that too at the start I had to take the readings so many times for surity. A bhaiyya in my neighborhood had much interest in religious activities but nothing was going right in life. Marriage delay, and earning issues I also knew that he is having but he wanted me to perform a complete Aura scan and verify that with his Birth Chart.

when I read the Horoscope, which I will mention here, Mars in sitting with Saturn in the 8th House, and also the Mahadasha of Mars was about to end. I told him this will show him the Malefic effects of Mars. Good point was that it makes the person brave! So I asked bhaiyya whether he faced frequent cuts, accidents, and injuries.

He came to meet me with some Guavas and a Thumbs Up bottle to greet me. I was also waiting and was ready to discuss the problems face-to-face. He affirmed the condition adding that last month only he was sitting on the motorbike and came before a cow and the rider was injured, he also fell but with minor scratches. He asked me how I analyzed. Then Bhaiyya showed me a Neelam stone and asked does this suit me. Many years ago a Pandit saw my Birth Chart and told me to wear this ring essentially so what is the positivity of this Neelam, does that suit him? It was a Natural Blue Sapphire and seemed to be of Ceylon origin but there were many inclusions in it and by which I doubted might not be fruitful for him.

I kept that stone on the mass cable and checked with the sample of Positive how much positive energy it had. The scanner showed a reading of 60%. Then I checked with Bhaiyya's aura. When he hold that Neelam in his hand the positivity of his aura decreased immediately and as soon as he left it the positivity came back to normal! So I told him no it does not suit you. If we see the chart in the traditional method, Saturn is sitting in the 9th House and is fruitful so the astrologer suggested he wear Neelam but when we see the Kundli in K.P then Saturn is sitting in the 8th house in the enemy house so it will cause a problem only instead of benefits. Then that Bhaiyya told me yes maybe that astrologer gave me the wrong opinion because when he wore that Neelam he faced huge losses, a court case came upon him and he had to sell his JCB machine also after that. I said,

"There were chances of imprisonment also"

He said, "It has already happened. On the auspicious day of the Holi festival I somehow ended up quarreling with a policeman and that fight fired up to physically mess up so I had to spend that night in jail. Just for one night but I pray that thing will never repeat in my lifetime. Tell me the remedies." After that, I suggested Pitra Dosh's solution and told him the Rudraksha rosary he is wearing doesn't suit his aura. After a few days, he told me yes he was sleeping peacefully after removing that rosary, and also after doing the pitra dosh remedy are positive changes in the family.

CHAPTER FOUR

POWER OF HANUMAN CHALISA

In some movies especially horror genre movies, they show that someone reads out the Hanuman Chalisa and the evil spirit goes away. I can't say anything about it, the movie is just for entertainment but yes I believe Mantras and Shlokas have special vibrations when we pronounce them and that is the reason why mantras are given so much importance in religious texts. Now leaving those things aside I would like to mention my experiments with mantras esp. The Hanuman Chalisa.

It was my initial days of Aura Scanning. I had just completed the training and bought it so was so excited to test it. From the videos, I learned that if we take the reading of the Positivity of a person, suppose a person's aura is 70% positive at that time then if he starts chanting a mantra continuously there will change in aura reading on this scanner. Either the positivity in his aura will increase or it will decrease. If that mantra does not go with his aura then the positivity of that person will decrease, else it will increase and expand. I had to test this somehow so I took the help of my Dad. He agreed to read out the Hanuman Chalisa aloud. That video I still have that with me any curious can take it from me! So the experiment was started.

For the experiment, I took a sample of planet Mars and took a reading of my father's aura. The Aura Scanner showed Mars is 50% positive for him. Then he started chanting the Hanuman Chalisa. I kept no changes and keep my hand holding the aura scanner at a still position. He was chanting it aloud and at first, there was a 10% increase in reading and no change till 30 seconds but suddenly it started to show more reading and went up to 100%. This means that chanting Hanuman Chalisa will give him the positivity of Mars planet to him.

Aura Scanner - 50% before saying Hanuman Chalisa

Aura Scanner - 90% while saying Hanuman Chalisa

I am the kind of person who takes this kind of thing lightly and doesn't believe easily that Mantras hold so much power but I was seeing it! What could I say I myself was experimenting. After that day I did this kind of experiment with many people. Once my relatives came to see me. They want to check what is this Aura Scanning thing so on my suggestion they all started saying RAAM... RAAM... RAAM... RAAM.... due to continuous chant the positivity in the environment must have increased so the aura scanner opened to maximum. Then suddenly he said what if we say negative things now will the positivity decrease again so I told them to say MARAA... MARAA.... MARAA.... MARAA... and yes the scanner started going less and less. They were all giggling and smiling what is this strange instrument you bought now.. how it functions..

CHAPTER FIVE

My Failures to read Correctly

"Astrologers can be wrong but not Astrology.

No one has come from the mother's womb well taught. We all are learning and getting updated. I have faced a situation where I took the reading and suggested to someone that Ketu's stone Lehsunia will suit him but the reverse happened. The same day he raised both hands saying he don't want to wear any stone. Whatever happens his ardent devotion to Lord Mahakaal is enough.

Firstly, Cat's Eye and Gomedh should never be worn with Gold because they are Paap Graha. Secondly, Gemstones intensify the results of a planet, when you start a car you feel a jerk, and if you stop the car at that time you can't drive. So if you wear a stone some changes may be visible still you should wear it for at least a few days to conclude. Not only this, but also I failed to answer people's True/False Questions correctly. It was because I was at my initial stages then. Anyone can ask me anything now, no issues. The machine is accurate. The theory is right. It is just a matter of practice and experience."

CHAPTER SIX

Aura of Water

One night while on my night duty, working from home, I got the idea that I should try performing Aura Scan on plants or any other object other than Humans. Suddenly I got an idea why not do this with water? So I went out of the room took a small amount of drinking water from a bottle and put that into the sample boxes and came inside my room. I kept the bottle on the window and started operating the Aura Scanner. The very next moment I was filled with joy and amazement both at the same time. How was it pointing toward the bottle? If I turned to the right and the bottle came to my left, it pinpointed to left, and if I turned right then it pinpointed toward the bottle on the right. It was a situation that was tough to handle for me. Am I holding something which is so sophisticated and accurate, I mean how...? So this set of experiments continued for days... I used samples of tap water, RO water, and Well-water. Aura Scan differentiates well between all different kinds of water. When I used RO water as a sample and tried to search for which direction the scanner is pointing it pointed to the drinking bottles and if I was away from those bottles then it pointed towards the Kent-RO itself. But when I used Well-water as a sample then the result was that it pointed toward the Well. I don't want to overpraise it but this is a very basic feature of Aura Scanner.

CHAPTER SEVEN

The Pishaach Yoga

If one understands his Planets, life becomes easier.

P.Y or Pishaach Yog is a certain kind of negativity that enters a person's aura due to the influence of a negative spirit. Negative frequencies of this kind are something which is least talked about but anyone can fall prey to it goes to an area that is under the influence of spirits then he will be caught by an Itar Yoni – Bhoot, Pret, Pishaach. We have heard stories of this from the mouth of our elderly people.

If a person is caught by any evil spirit then his Aura becomes up to 70% negative. If anyone performs Aura Scan on that person then the result will be that the Scanner will open up and show 100% result while taking a reading with the Pishaach Yog sample. There is a relative of mine who had complained of financial issues and almost every big endeavor in his life at that time was at a stay. No progress. He approached someone on youtube for an aura scan. That Pandit Ji declared that everything is ok except for Shani and Rahu so he has to perform a havan for the peace of planet Rahu. He paid that Pandit Ji aura Scanning fee of 2,100/- and took the 'helpful video'. Pandit Ji had also suggested to him that wearing a Ruby in Gold will be very fruitful so he did that also in presence of me.

After 2 months I got a call from him saying that no difference. I asked him if should I perform Aura Scanning on him through the photo. He said to do anything but it should work.

So as per protocol, I checked positivity – 70% and then Negativity was also 75%. Did I think how this much negativity in his aura? (If someone's negativity is more than 40% then it means there is some major problem and we have to check every Doshas sample for him)

Nazar Dosh was not detected.
Bandhan Dosh was not detected.

I thought Pischaach Yog may not possible but then also I had to test it after all. When I took the reading I was shocked because the scanner opened from 0 to 100% which means that person was influenced by a paranormal entity. I didn't understand what to do because I was shooting my activities to show him later so I just resumed the readings without showing any reaction.

I sent that video of the reading to my teacher who told me that my analysis is correct and not only that, it also means that this spirit is not an old 20-30 yrs one but it is only 4-5 yrs old.

I told my mom and she said,

"Surely it is possible. He had engaged a land 4 yrs from now, which was said to be cursed. Everyone stopped him to purchase that land but he was unwilling to listen to anyone. Neither did he perform any puja on that land. So now he might be suffering because of that".

Now I had no other option but to inform that bhaiyya about the reading and then he also agreed and said,

"How you found out that Pandit Ji's aura scanning couldn't find it?"

It was because my Aura Scanner is of the latest technology and more effective than other people's machines. Pandit Ji's scanner is of 10 yrs old technology that didn't have a sample of P.Y. that's why he missed the real cause of the problems.

Thankfully the cause of the misfortunes was found at the right time. My apologies to Pandit Ji I am not taking anyone's name here.

CHAPTER EIGHT

How I came to Know about Aura Scanner

I read a book during the lockdown period. It was about how a Yogi saw the aura of a person and told him which disease will occur to him in the coming days. There was information like which color of aura denoted what kind of disease. So I searched for it on the internet. There came some images in which portraits of people were shown with different circular and irregular color formations around their bodies. It was called Aura Imaging. Then on youtube, I came across a video titled Aura Scanning and it rose my curiosity. I ended up contacting that Pandit Ji and paid him a fee of 2,100/- to know about my planets, Chakras, and which Gemstone will help me to rise in life. Sad to announce but that scanning process was inaccurate which I came to know later on when I got trained myself. The problem is how will you verify the results. Someone will say your aura is 75% positive, someone will say your aura is 25% positive and someone might say your aura is 100%. How will you verify? What I do is follow whichever is most logical.

Two years later, I searched for more videos and I found an astrologer who had uploaded hundreds of videos on aura scanning.

His name is Pt. Rajendra Prasad. Very polite person! I paid him also to get the same analysis, esp. profession, and marriage. He said 9 months later promotion, and 1.5 yrs later marriage. My increment happened suddenly just 5 months from that date and the marriage prediction time has not come yet. Readers can decide about the analysis but in my opinion, this prediction too is inaccurate. He told me that my Chakras are all fine, no Pitra Dosh, no Kal Sarp, no Bandhan Dosh. Venus is bad so do 10,000 chants of the Shukra mantra and wear Opal. I only did 1,000 Mantra jap for 10 days and I felt happy inside, and more positive. I thanked him on WhatsApp. What I thought is that this is a good business you tell everyone planets, chakras, and some questions and earn 2,100 from one person! Practically all work is done by the scanner and you get 2100 benefits with no investment, just having an aura scanner is enough. Life set!

But Destiny has its plan. I didn't get the Aura Scanning device from him and neither did I get involved in earning a lot of fees from people, except a minimum amount for now.

A person was seen in a seminar video, aura scanning the local newspaper to tell which person is dead in the newspaper and which is alive. I thought let me check the video I had a lot of vella time for these things at that time.

There are 121 videos on the channel I watched it all because someone was talking with logic as well as a demonstration. There is a famous saying in our ancient texts "Pratyaksham Kim Pramaanam" if something is in front of the eyes, the proof is not needed. So this was the place from where I got training to operate Aura Scanner. Initially, there were many failures! I am still learning!

CHAPTER NINE

How to find Lost Object at Home

In the training on how to operate the aura scanner, it is taught how to find a lost object in the house with the help of the scanner. Suppose that you have a habit of getting your Wallet so for that you just have to create a customized sample of your wallet and whenever your wallet will be lost this aura scanner will lead you to the wallet. You have to insert the sample of your wallet inside the Aura Scanner and when you raise it and wait for the reading it will point in the direction of the wallet. Then when you reach that point it will fully open. For eg., if the wallet is under your bed when you put this scanner on top of the bed it will fully open which is the signal that the wallet is under that portion of the bed.

In case, you can't believe this and you have the aura Scanner with you I tell you how to test it. I have also uploaded a video on my 'Singhaasan Battisi' youtube channel. What you have to do is - Take 3 bowls and fill each of them with three different edible items. Fill the first bowl with Rice, the second one with Tea, and the third one with Wheat, as shown in the picture.

Episode - 05, Singhaasan Battisi

then put the sample of Tea in the empty sample box and insert the samples in the scanner. Now turn on the scanner and take the reading. Surely after a few seconds, the scanner will point toward the Tea bowl. Keep it near the Wheat bowl then also it will pot to the Tea bowl or keep it in front of the Wheat bowl, then also it will keep pointing to the Tea bowl itself. The scanner will point toward the object you choose to take as a sample.

Now bring the scanner above the Rice bowl, it will not open. Bring the scanner above the Wheat bowl it will not open. Bring it over and above the Tea bowl, which we have chosen as our sample, you will be filled with happiness to see that it opens all the time you do so.

So in this way, we can find the lost object in the house. To learn and master this technique you can do the same experiment with any edible substance from the kitchen. why I advise you to choose edible objects because then you will not have to bring the photo printout and use it in the sample. Just putting some amount of edibles inside the sample box is enough.

You can take some tea leaves in a bowl and ask someone to hide that bowl without informing you. The hidden tea should be kept in a place where it is uncovered such as on the balcony, or below the table (not in the drawer of the table). The same tea leaves you have to take as samples and then you can perform the experiment of finding it. The method will be the same as mentioned above just we have replaced the substance. Even this topic's demo is uploaded on my youtube for a better understanding of people who are curious to experiment.

this episode is uploaded on the channel Singhaasan Battisi

About Aura Scanner

CHAPTER TEN

AURA SCANNING - PURE SCIENCE

Aura Scanner is a scientific device. It is a battery-operated device. This machine gives us the answer either in Yes or No or in percentage, depending upon our question. For different question different sample is used. Take the example of a T.V remote used to change the channels. The remote is the receiver, the sample we put in this is the channel and the T.V screen is the result that we obtain from this Aura Scanner. Without choosing the correct channel we will not be able to watch the correct show, the same way a particular sample is used in the scanner to obtain a result to our question.

Sample

What is a sample and what is it made of?

A sample is a small circular box that consists of printed Symbols. For example, the sample of CHILD BIRTH is having a symbol of a newborn child, the MARRIAGE YOG sample is having the symbol of a couple taking circumambulations of Marriage, the sample of 1 Mukhi Rudraksha is having a printed image of 1 Mukhi. This scanning device needs input to know which kind of result are we looking for.

There are a total of 400 samples in 3 different Sample Boxes, which come along with this Aura Scanner. I have explained how to use each sample in detail in the respective chapters.

What is the application of the sample?
Suppose, a Jataka comes to us and asks about the misfortunes he is facing in life. Maybe his time is not running well and he is not able to understand the reason behind it. For that first of all, we check with the sample of Positivity. The positivity in his Aura should at least be more than 40%. Then we take the sample of Negativity and take a reading. The Negativity will be more than 40% for sure. What can be the reason for Negativity in Aura?

1. Due to the presence of any negative object in the house.
2. Due to bad habits like smoking, drinking, and non-veg food.
3. Negative thoughts or speaking badly about others all the time. Like some people always find bad about others in everything. While on the other hand, some will speak always good and give blessings to others. So this makes a lot of difference.

In this way, by using the sample of Negativity only we can come to know about the root cause of the problem. I once bought a metal Bracelet from Amazon. It looked cool so I ordered it and wore it. After a few days, I felt all my work getting delayed. Payments were pending. Suddenly one day I thought I should check the positivity of this Kadaa. It was least positive and highly negative. So in this way many times these samples will guide you in all phases of life. Maybe you are about to purchase land. Just bring some soil from that land and use the sample of Positivity, Negativity, Bandhan Dosh, P.Y. Maybe someone had passed away on that land or an animal was buried there! You will come to know whether you should purchase it or not. Whether after purchase that land will be profitable to you or not can be known by giving that command to the scanner using the sample of Positive.

CHAPTER ELEVEN

Uses of Aura Scanner

General Use

All living beings on Earth can use this scanner for some or the other use. Maybe someone's marriage is delayed or he is not getting a job maybe someone is not having a child after many years of marriage. We can get answers to any question and not only that but also suggest the remedy accordingly. Maybe someone has Negativity of more than 40% and that's why there is a kind of blockage in his Aura. Once he does remedy it he will see that the answer in the machine will also come as Yes and also he will be getting desired results in his life.

Choosing the correct Life Partner
Which Business will be profitable
Whether to purchase Land or not
Which planet's power should be increased/decreased
Health Check
Vastu Check and Remedies
Which location to do Water Boring

Suppose you are about to buy land or property or a flat. Maybe someone buried a dead animal there and you are unaware of it, this will easily come out in an aura scan. The Vastu of a flat will also reveal with one Vastu visit to that place. If the flat is having minor Vastu Dpsha then go for it! purchase that land and it will only be fruitful to you. There are actual examples in which people have found pig's tooth, which is the most negative substance, from buried underground with taking help of this device!

Talking about the share market and whether one can use this for getting a perfect prediction of which share will rise or fall, I don't know. The reason behind it is how will you scan the aura of a share I don't know but after reading this book maybe someone might approach me with some suggestions. Anyone is welcome. Whether you should do share marketing or not, we can find out easily from this. Someone who has got favorable Rahu will mostly get positive results in share markets. What occupation one should choose, we can find out. Should he go for a job or business, we can find out. How much profits he will earn in business we can find out in percentage! I think this is enough to give you an insight.

Aura Scanner was invented to find out Geopathic Stress and was later used for all kinds of Vastu checks then when it came to India it was used for checking planets and other uses. Later on, many other samples were invented to use for checking Gemstone compatibility, check Marriage Yog, Santaan Yog, and even Divorce Yog. Suppose you tend to forget you kept the wallet at home. Then you can find it out with the help of an aura scanner. For that, you have to make a custom sample of your wallet. When you insert the sample of the wallet in the scanner it will point out the direction and lead you to that place! Same way lost Gold/jewelry piece, watch, keys, or mobile phone can be found at home. A lost person is dead or alive, we can know from that person's photograph. However, to know where is his location we have to suggest options to this scanner then only we can get answers. This particular subject I am still trying to learn and not got perfection yet.

CHAPTER TWELVE

How to Aura Scan?

For Aura Scanning either you have to have the person's photograph with you or he should be present physically before you or his Sliva DNA must be with you. If the person is present physically you don't have to attach the mass cable just raise the Aura Scanner, switch it ON and keep the Scanner at an angle of 85° from the front and the scanner will give you the reading. It is as simple as that.

Also, you have to stand/sit barefooted so that your feet touch the ground and this scanner gets Earth element from the ground. Your reading will be quicker. If you should be getting a reading in 30 seconds you will get that reading in 20 seconds. If the person whom you are scanning is also barefooted then it is much better.

The second scenario is when the person is not present before you. In this case, you need to get that person's photo printed in Postcard size. Any photo in which he is standing alone and no one is standing behind him can be used. The photo can be Five years old or even ten years old, it won't matter if the reading will be of the present day. Now attach the mass cable with the scanner. Two wires and two circular aluminum mass is given to you just attach them with the cable wire. Put this mass cable on the person whose aura you have to scan and then you can perform the readings in the same way.

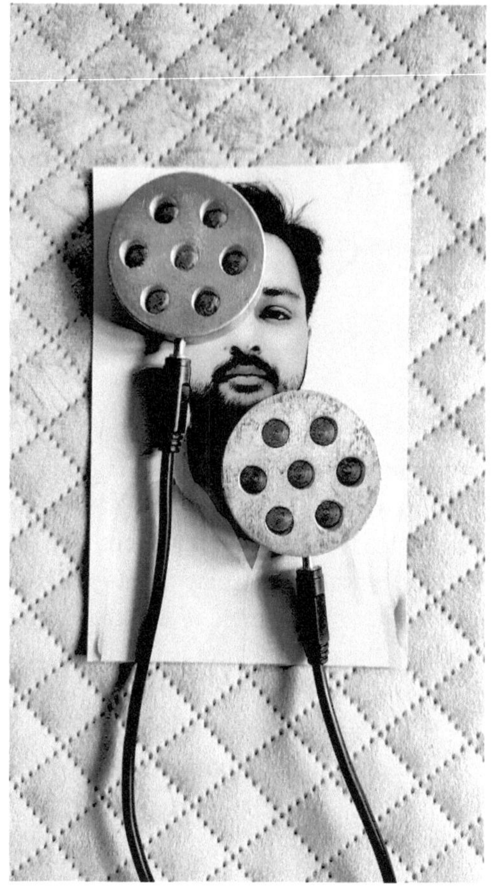

Mass Cable kept on photo

The reading should be taken by keeping the scanner at an angle of 85° and not 90°:

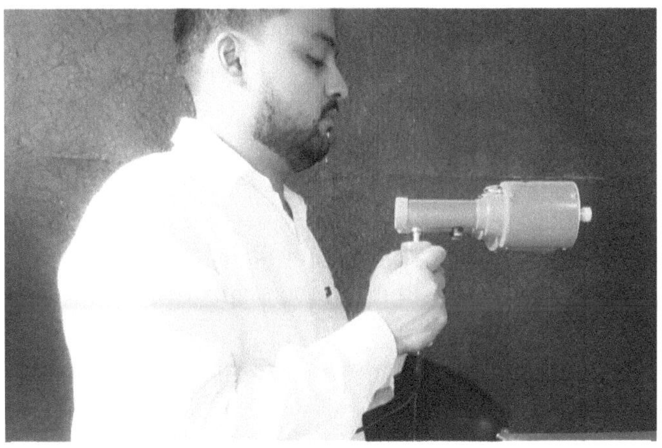

This is a 90° angle which is the wrong way

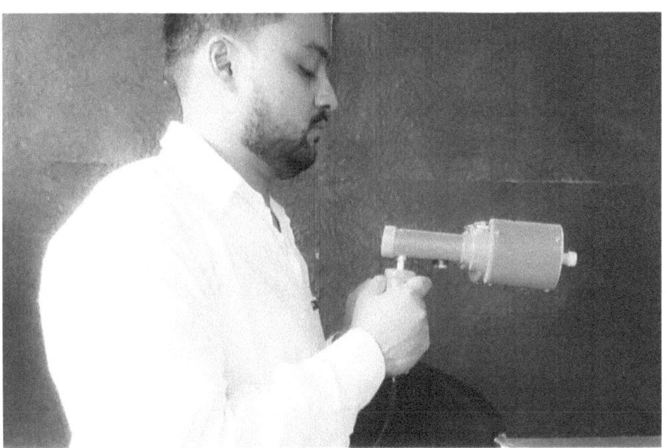

This is an 85° angle which is the correct way

As you can see clearly my hands are also only at the handle, we don't have to touch the top portion of the moving area, or else you will disturb the reading. The area which is left free to rotate will show you the reading.

You can take the reading while sitting or standing. Time can be night or day. Direction can be any. There is no other restriction other than as mentioned above.

Reading is understood from the space between both limbs of the scanner.
If the space between them is 180° the answer is 100%.
If the space between the limbs of the scanner is 90° the answer is 50%.
If the limbs are joined together with no space between them then the answer is 0%.

0° meaning the answer is NO.

50% answer.

180° or 100% the answer is YES.

In Yes/No questions such as Pitra Dosh, Manglik Dosh, and Bandhan Dosh, the answer will come in Yes/No. If the angle is 0° that means the answer is No. Otherwise, if the angle is 180° it means 100%, and the answer is Yes.

Readers must watch the video from youtube for better understanding although I have already covered all the things that are to be kept in mind. If he learns how to hold this properly then he can get answers to any question.

There are some prerequisites of Aur Scanning which I have mentioned but still, but I will again list them in a separate chapter for reference.

CHAPTER THIRTEEN

SCIENCE BEHIND AURA

What is Aura?

Aura is an Electromagnetic field of Energy that is present inside every human being from his birth till death even after death the body is not active but the Aura is still there. Dead bodies have a highly Negative Aura with 0% positivity. If anyone provides me with a photograph without revealing whether that person is dead or alive then also I can find out with the help of this scanner because dead people's photographs will show 0% positivity. In hindi, Aura is called 'DivyaKanti Valay' or 'Aabha Mandal'. Our Saints were picturized showing a round figure around the head which can be noticed in paintings of Guru Nanak Dev Ji and Guru Govind Singh Ji and a few others. This is Aura. Saints have a Golden Aura. No one can see the Aura and Wikipedia says there is nothing called aura but scientists have invented the technology to catch the frequencies of Aura. Someone who can read Auras can tell you which disease you are having from the state of your Aura. This is not magic but is having logic behind it.

Aura is not a part of the physical body but it exists in the form of an Energy Field around the physical body. This is also called the Energy Body or Praan Shareera. Electricity can not be seen or picturized but it can be felt. In the same way, the subtle body and its frequencies are not visible to the naked eye but through this device - Aura Scanner, you can know the state of your Aura. Aura Scanner will give you first receive the frequencies of the person's Aura and gives us the result either in Yes/No or in percentage. Some answers will be in Yes/No while others will be in percentage.

Plants too have an Aura, animals too have an Aura so do all non-living things. But here in this book, my subject is how to study the Aura of Human Beings and Plants. Children have a softer Aura while adults have a strong Aura so it is advised that an Aura report should be made for adults only.

Aura Photography

"Science has proved that our aura has the power to determine our health in the future."

- P.M Narendra Modi speech

In 1939, Russian inventor Semyon Davidovich Kirlian invented a method of taking portraits using a metal plate and an electric current that he claimed could reveal the “Lifeforce” of his subjects that even predicts illnesses. Kirlian photography is regarded as the predecessor to what we now know as aura photography. In the 1980s, a man named Guy Coggins invented the aura camera for the mass market: the AuraCam 3000. Coggins later introduced the AuraCam 6000, which is what most aura photographers use today. It is a pioneer independent research work in the field of Aura Reading. Kirlian photography has been the subject of scientific research, parapsychology research, and art. Paranormal claims have been made about Kirlian photography, but these claims are rejected by the scientific community. To a large extent, it has been used in alternative medicine research.

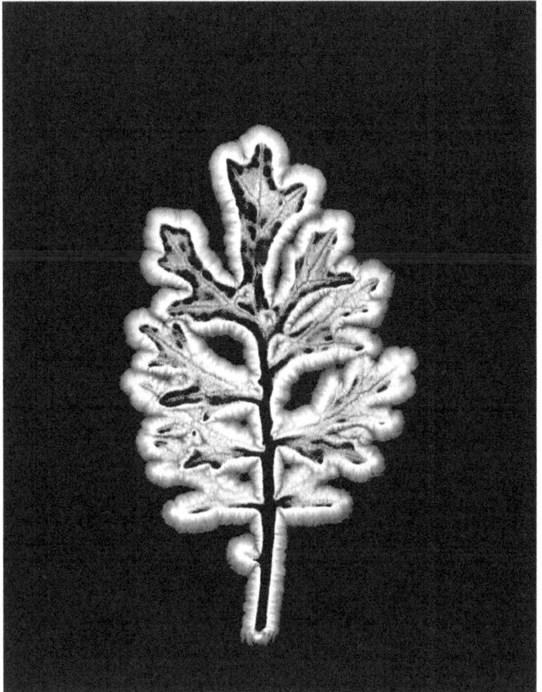

Kirlian photograph of a dusty leaf

Kirlian believed that images created by Kirlian photography might depict a conjectural energy field, or aura, thought, by some, to surround living things. Kirlian and his wife were convinced that their images showed a life force or energy field that reflected the physical and emotional states of their living subjects. They thought that these images could be used to diagnose illnesses. In 1961, they published their first article on the subject in the Russian Journal of Scientific and Applied Photography. Kirlian's claims were embraced by energy treatment practitioners but his theory was rejected by the scientists.

Biomagnetism

Biomagnetism is the phenomenon of magnetic fields produced by living organisms; it is a subset of bioelectromagnetism. In contrast, organisms' use of magnetism in navigation is magnetoception and the study of the magnetic fields' effects on organisms is magneto biology. (The word biomagnetism has also been used loosely to include magneto biology, further encompassing almost any combination of the words magnetism, cosmology, and biology, such as "magnetoastrobiology".)
- Wikipedia

Several animals are suspected to have the ability to sense electromagnetic fields; for example, several aquatic animals have structures pote capable of sensing changes in voltage caused by a changing magnetic field, while migratory birds are thought to use magnetoreception in navigation.

Birds observing Bio-Magnetism

Dowsing

Ancient Dowsing Tool

Human beings are God's most beautiful creations. They have been gifted with numerous powers and energies. Nobody has ever scaled the depth of the human mind. We, as humans, have not even utilized our limitless powers.

One of the modern concepts of Vastu Shastra is 'Dowsing' which is only 20% accurate. Dowsing is an art through which we can find the root cause of our problems. With the help of this Celtic art, we can find simple, appropriate, and possible solutions to solve our problems. This practice has been traditionally used for divination.

The literal meaning of 'dowsing' is to dive after going deep. In the practice of Dowsing, a 'pendulum' is used. A weighted object that hangs from a long thread or a chain is called a pendulum. The pendulum dowsing technique is being traced in the land with the search for treasure, water well, landmines, gemstones, and other mineral substances for centuries.

I just wanted to mention that this famous technique also is only 10-20% which I have self-experienced but Aura Scanner is upto 100% accurate if performed correctly.

About Aura Scanner

To measure the Life force, Aura Energy, a device called Aura Scanner was invented. This was invented by scientists in Germany. There was a city built by the government which had buildings of the same design. People were affected by many accidents in a particular area. So the government appointed a group of scientists to find out the reason behind this secret. They concluded that there is a certain Negative Energy coming from The Earth on that road which was the root cause of accidents. Thus this aura scanner was invented and later on became popular and came to India. We use it for scanning our planets, Chakras and to know about a person's job, business, and marriage. This is a boon to society if utilized properly.

To measure the Life Force in a body, Aura Scanner is used. We can measure aura and help people tell him about his weak & strong planets. If the planet's reading in Aura Scan comes to less than 40% that means that the planet is weak in his birth chart. As a remedy, he must donate that planet's object. On the other hand, if the planet's reading is more than 40% that means that the planet is lucky for him, In most cases, those gemstones will also suit him the process I have explained in the coming chapters.

An aura scanner is an electronic device that runs on a battery. It comes in two pieces. It has also got 2 wires, 2 mass cables, and 400 samples. What we have to do is hold this scanner in front of the person whose aura we want to check. Wire connection is not required to scan someone present before you. If a person is away from you then you can scan his photograph to get the same results. The accuracy is the same with all the 3 types of scanning. The hand has to be placed on the handle only and the upper portion is left free to move. It has got ball bearings inside it, to allow it to move freely. When you turn it On by putting power On also, it will rotate 360 degrees. So, KINDLY TAKE READINGS ONLY AFTER PROPER TRAINING FROM AN EXPERT.

CHAPTER FOURTEEN

Pre-Requisites for Aura Scanning

When we buy a new brand's mobile it may take some time from our end to operate all its features. It may take hours or maybe days when we get used to it. The same goes for the case of Aura Scanner. Once you buy it, you must necessarily take part in the two days Zoom workshop to learn all the minute details to operate this machine. There are people who bought the scanner but are still not able to operate it, and that's the reason I need to bring out this book. Here I will mention the most important 3 points to be kept in mind after which you can operate the scanner:

- At least 4 chakras of you must be balanced then only you will be able to operate it properly. If the chakras are imbalanced you can ask someone to check them for you and if it is unbalanced then you may have to use the essential oils of the respective chakras, as per the guidance. After usage when 4 Chakras are balanced you will automatically start getting proper results.
- You have to keep the Aura Scanner at an angle of 85° from the front and not at 90°. If you keep that at an angle of 90° it will rotate on its own because of the ball bearing present in it, thus the reading will be wrong. For this, you can refer to the picture in the next chapter.

- While scanning you must be sitting with no slippers or shoes. This aura scanner should get Earth element from you to give the results quickly. If the Jataka also sits with naked feet then nothing can be better.

To scan someone's aura you should have any 1 of the following things about that person-

a. the Jataka himself is present before you
b. Jataka's Sliva sample/DNA
c. The photo of Jataka

I have mostly used photographs for 98% of the people whom I have suggested, so around 300 photo printouts are now here with me!

- This photo should be unedited, the person should be alone in the photograph and even in the background, nobody should be present. Taking a quick selfie is the best way.
- Photo should be of postcard and in the form of hardcopy. Only printed photos will serve the purpose. Merely opening it on mobile will not do because mobile screen change frequency rapidly and laptop and PC has radiation in it. Some users do fake readings by scanning people directly from a laptop screen. This method is wrong.
- An old photograph from 5 yrs ago will also show the results of his present Aura. some people think that the latest pic, without wearing any gemstone, will only give correct results - it's not true! The reason behind this is that you might be scanning a person's photograph who has passed away now. But the photo must have been taken when he was alive! That means that the reading in Aura Scan will show the results of the present situation only and yes you can also ask the scanner questions from the past!

Sample Box I

CHAPTER FIFTEEN

BANDHAN DOSH

Bandhan Dosh is caused when a person does black magic on someone due to jealousy. It is a kind of Tantrik kriya to cause negativity in his life. Bandhan Dosh on a person is present or not, can be checked with the Aura Scanner by using the sample of Bandhan Dosh or we can call it Tantra Badha Kiya Karaya. If the Negative Aura of someone comes to more than 40% then it is necessary to check his Bandhan Dosha and when we put a sample of Bandhan Dosh in the aura scanner it will open to 100%. This reading will either come in 0% or 100% because it is a Yes/No question - either it is present or not present so finding it out is easy. After this, we can also find out who did it by giving the scanner commands.

The cause of Bandhan Dosh

Sometimes a boy likes a girl but the girl rejects him then he might do BnadhanDosh other with the help of someone he will make her eat some special kind of charming food. Then a few days later that girl will be affected. Anyone having jealousy also does this extremely bad activity.

Effects of Bandhan Dosh

If Nazar Dosh is found in someone's Aura it decreases the productivity of that person by up to 75%. He gets a loss of Wealth and Health of up to 75%. What I mean to say is that when someone catches a Nazar Dosha, his earnings will fall by 75%. He may be earning lakhs and saving one Lakh per month but after Nazar Dosh, his savings might decrease and he would be able to save only an amount of Seventy Five Thousand. This is not a luck factor and is neither a mystery now. The reason can be Nazar Dosh in his Aura. First Nazar Dosha affects the finance of a person and then it comes to his Health. Health issues are found which leads to other problems simultaneously.

Remedy for Bandhan Dosh

If someone is found with Nazar Dosh, the easiest remedy is to get Aura Booster from us and keep it always with you, esp. when going out for some work. Also, you will get this Aura Booster online through different sources but I would suggest just checking the positivity of our Aura Booster and comparing it with other manufacturers, you will observe a huge difference of a minimum of 20%. By using our Aura Booster the positivity increases by 40% which is a game changer in career path and all. Our aura Booster has a life of more than five years which means the day you start using our aura booster it will be effective for you for Five years and more. The negative frequency of Nazar Dosh U.V- minus will not affect that person till the time he is in contact with our Aura Booster.

How to check Bandhan Dosh

You can follow the following steps to check Bandhan Dosh for a person. If you are away from that person just get his photo and perform an Aura Scan on his photo printout. **Directly performing Aura Scan on his pic opened on mobile should be strictly avoided**

because the mobile screen's frequency changes rapidly and the reading taken from that mobile screen will not give correct results. The following steps are followed to check Bandhan Dosh in anyone's aura :

1. Put the Mass Cable of the Aura Scanner on the person's photo printout or if that person is in front of you then no photograph is required and no mass cable needs to be connected. You can just scan him by standing in front of him and taking the reading as shown in the pictures.
2. Place both samples of **Bandhan Dosh** on both limbs of the Aura Scanner.
3. Switch ON the Aura Scanner.
4. Place both the limbs of the Aura Scanner at an 85° angle at the front.
5. Wait for the reading and observe the result. It will indicate either a Yes or No. If the scanner opens making an angle of 180° / 100% then that means the answer is Yes. The person is having Bandhan Dosha. On the other hand, if the scanner does not open at all that means No Bandhan Dosh in his Aura.

CHAPTER SIXTEEN

NAZAR DOSH

Nazar Dosh is checked with the Aura Scanner by using the sample of U.V- or we call it UltraViolet minus. If the Negative Aura of someone comes to more than 40% then it is a necessity to check his U.V- Some people, especially females and newborn babies are prone to acquiring Nazar Dosh. If there is Nazar Dosh, getting a solution for planets won't help. Every problem can be overcome only when you come to know what the problem is. I have listed here the cause and remedy for Nazar Dosh.

Causes of Nazar Dosh

People who have 32 teeth or someone with jealous intentions in mind say something then this produces a kind of Negativity which is having the frequency of Ultraviolet negative. So that's why in today's world we need to be more careful. You might have heard people saying that they easily fall prey to Nazar Dosh.

Effects of Nazar Dosh

If Nazar Dosh is found in someone's Aura it decreases the productivity of that person by up to 25%. He gets lots of Wealth and Health to up to 25%. What I mean to say is that when someone catches a Nazar Dosha, his earnings will fall by 25%. He may be earning lakhs and saving one Lakh per month but after Nazar Dosh, his savings might decrease and he would be able to save only an amount of Seventy Five Thousand. This is not a luck factor and is neither a mystery now. The reason can be Nazar Dosh in his Aura. First Nazar Dosha affects the finance of a person and then it comes to his Health. Health issues are found which leads to other problems simultaneously.

Remedy for Nazar Dosh

If someone is found with Nazar Dosh, the easiest remedy is to get Aura Booster from us and keep it always with you, esp. when going out for some work. Also, you will get this Aura Booster online through different sources but I would suggest just checking the positivity of our Aura Booster and comparing it with other manufacturers, you will observe a huge difference of a minimum of 20%. By using our Aura Booster the positivity increases by 40% which is a game changer in career path and all. Our aura Booster has a life of more than five years which means the day you start using our aura booster it will be effective for you for Five years and more. The negative frequency of Nazar Dosh U.V- minus will not affect that person till the time he is in contact with our Aura Booster.

How to check Nazar Dosh

You can follow the following steps to check Nazar Dosh for a person. If you are away from that person just get his photo and perform an Aura Scan on his photo printout. directly performing Aura Scan on his pic opened on mobile should be strictly avoided because the mobile screen's frequency changes rapidly and the reading taken from that mobile screen will not give correct results. The following steps are followed to check Nazar Dosh in anyone's aura :

1. Put the Mass Cable of the Aura Scanner on the person's photo printout.
2. Place both the samples of **U.V-** on both the limbs of the Aura Scanner.
3. Switch ON the Aura Scanner.
4. Place both the limbs of the Aura Scanner at an 85° angle at the front.
5. Wait for the reading and observe the result. It will indicate either a Yes or No. If the scanner opens making an angle of 180° / 100% then that means the answer is Yes. That person is having Nazar Dosha. On the other hand, if the scanner does not open at all that means No Nazar Dosh in his Aura.

CHAPTER SEVENTEEN

Pitra Dosh

What is Pitra Dosh?

Pitra Yog and Pitra Dosh are two different aspects of Kundli. Pitra Dosh is a kind of negativity that comes to us if we fail to perform the proper ceremonial duties of any deceased ancestors. In hindi we call it Shraadh. Pitra Dosh will affect the coming generations more. Whoever performs the ceremonial rights, if it is not done properly then he and more than he, his coming generations will suffer from Pitra Dosha.

Symptoms of Pitra Dosh

- If the Pitra Dosha is two to three generations old then there will be fewer male no. of child in family, female children will be more.
- The children will acquire spectacles at a small age.
- If there is a male child, he will not listen to the parents.
- Unwanted quarrels among the family members.

These are some of the most important symptoms of Pitra Dosh which will surely be present. We can check with the sample of Pitra

Dosh if the scanner opens in Pitra Dosh then we can confirm with the person about these symptoms. Pitra Dosh is a very important aspect of one's life and should surely be checked for every person.

Remedies of Pitra Dosh

First of all, let us see how to check Pitra Dosh, and then we will focus on how to deal with it. For that, I am listing everything step by step.

Insert the sample of Pitra Dosh in the Aura Scanner. The scanner will show the result as Yes/No. If Pitra Dosh will be there, the scanner will open to 100% otherwise if it is not there the scanner will not open at all. It will not open in percentage because this is a Yes or No question. Either Pitra Dosh is present there or it is not. There is no in-between.

If Pitra Dosh is not there then no further check is required. The condition is auspicious so we can check with other topics. But if Pitra Dosh is present, first we have to give the command to the scanner one by one that how many person Pitra Dosh is there. You can ask,

"Does Gaurav have Pitra Dosh of 1 person?"

Like this, we have to find out the exact no. of ancestors whose Pitra Dosh has occurred. Let's say there is Pitra Dosh of 3 people on someone. Next, ask:

"Does Gaurav have Pitra Dosh of 1 female?"

In this way, we have to find out exactly how many Male ancestors and Female ancestors are involved. We will get correct answers, even if we cross-examine. The next step is to confirm the name of our ancestor whose Pita Dosh is there like:

"Does Gaurav have Pitra Dosh of his par-dadi ji Kamla Devi?"

The scanner will indicate Yes or No. Once we find out exactly whose Pitra Dosh is affecting the person then we can check with the remedies. The same remedy will not apply to every Pitra Dosh. There are a total of 3 types of remedies. Any one of these will be the cure:

- Roti daan
- Sharaadh Kram
- Shayya Daan

We can check with the scanner which Remedy will work for which Pitra Dosh like to ask,

"Will Gaurav get dadaji's Pitra Dosh cleared by Roti Daan?"
"Will Gaurav get dadaji's Pitra Dosh cleared by donating Roti for 1 day?"
"Will Gaurav get dadaji's Pitra Dosh cleared by donating Roti for 11 days?"

If you get the answers cleared then for the next person also we can check in the other way,

"Will Gaurav get dadaji's Pitra Dosh cleared by offering puja in Gaya?"
"Will Gaurav get dadaji's Pitra Dosh cleared by offering puja in Badrinath?"
"Will Gaurav get dadaji's Pitra Dosh cleared by offering puja at Haridwar?"

1. Method of Roti Daan

This is the first kind of remedy and also the simplest one so you should ask for it first. For this, you have to donate some Rotis and food items that are cooked that day at home. Just keep those things on a plate and keep them outside the house in a lonely area. After that, the food might be taken away by cows or birds, or insects which is not a problem. Just that you take out food from the house for Pitras it is enough. Do it continuously for that number of days as it was answered and once you perform it if you re-check with the scanner Pitra Dosh will be gone.

2. Shraadh Puja

When questioning with the scanner if the answer comes that Pitra Dosh's remedy for an ancestor is Shraadh Puja then you need to find out exactly which place Shraadh Puja is required. It may be in the same town in a mandir or at Gaya, Badrinath, or Haridwar. You should strictly follow the guidance. Once you perform it if you re-check with the scanner Pitra Dosh will be gone.

3. Shayya Daan

This is the third type of Remedy and if the answer comes that Shayya Daan is required then the Brahmin will help perform all the rights. All the important items of living are purchased and properly donated to the Brahmin. Bed, Utensils, Clothes, Black Leather Shoes, Black Umbrella, etc. must be donated. After all this, when the Pitra Dosh will be re-checked it will not be detected.

CHAPTER EIGHTEEN

Job / Business Success

Suppose there is a boy who has just passed college and is about to start his professional career. we can check and tell him which path will be best for him - Job/Business. If Job, which Job? If Business, which Business? There is a separate sample for Job Success as well as for Business Success. It will give results in percentage. If the Job Success percentage comes to 60% and the business success percentage comes to 100% then he should go for Business, he will definitely earn more profits.

Next, we can also check which job/business type he should choose for the most profit. For this, we have to give commands,

"How much percentage will Healthcare sector jobs be profitable for Gaurav?"

The answer will come in percentage. Someone who is already working can verify this easily. Suppose that we want to check which type of Business will be more profitable, we have to question with the Aura Scanner one by one,

"How much percent will Tea Business be profitable for Gaurav?"

CHAPTER NINETEEN

MARRIAGE YOG

Our first question is whether the person has Marriage Yoga or not, whether he will get married in his life or not. If yes, when?

Take the person's photograph and put both the mass cable on it.
Insert the MARRIAGE YOG sample in both scanners. Turn it On and wait for the reading.
If the scanner opens, Marriage Yog is present and he will get married in his life otherwise if the scanner does not open then the Marriage Yog is not there.

If the answer is Yes then we need to find out the time of Marriage, and for that, we have to speak the question one by one. If we put the command, "Will Gaurav get married within 1 year?"
If the answer is Yes then we can stop otherwise speak the question for 2 years and so on.

CHAPTER TWENTY

WILL YOU HAVE YOUR OWN HOME?

Sample of Own Home

Aura has answers to any question which is related to us. Now, this topic is whether a person will be able to get his own house or not in the future. Sometimes a person comes to test us whether our Aura Scan is correct or not so he may ask "Tell me about my financial status." or he may ask, **"Will I be able to build my own house in this life period?"** Aura Scanner will not only tell the answer in Yes/No, but also it will tell us whether that property will be totally in his name or will it be a shared property with someone else's name share also or whether he will spend his whole life living on a rented house.

The sample of Own Home is used for this. The steps are listed as follows -

1. Put the Mass Cable of the Aura Scanner on the person's photo printout.
2. Place both the samples of **Own Home** on both the limbs of the Aura Scanner.
3. Switch ON the Aura Scanner.

4. Place both the limbs of the Aura Scanner at an 85° angle at the front.
5. Wait for the reading and observe the result. If the Aur Scanner does not open that means he will not have his own house in his life. If the scanner opens to 50% that means the person will have a property on a shared basis. He may be having his property in his wife's name shared with himself. Or if the answer is 100% then that is to be concluded as he will own property, Own Home which will be totally in his name!

CHAPTER TWENTY-ONE

Foreign Tour / Settle

Everyone is anxious to know whether in their lifetime they will be able to visit foreign lands. The Aura Scanner gives a clear indication of whether that will happen or not. Kundli analysis can be complex and maybe there might not be any specific Yod formation in the birth chart so the person might feel demoralized that this event will not happen in his life but with the help of the aura scanner he will be able to get a clear indication that he will surely go abroad. for this, we have to insert the sample of VISA and if the scanner fully opens then will go to another country otherwise not.

For Foreign Settle, you must use the sample of FOREIGN SETTLE from sample Box 1. You have to take a reading from the aura scanner as is already explained in the previous chapters. If the scanner fully opens to 100% then the answer is positive he will visit a foreign land.

CHAPTER TWENTY-TWO

How to check power of Gemstone

Before starting the Chapter on how to check Gemstone Compatibility, let's deal with checking the power of a Gemstone.

Mass Cable joined and Gemstone kept on it

1. Put the Mass cable one on top of the other.
2. Place the Gemstone on the Mass cable.
3. Turn on the Aura Scanner and insert in it the sample of Positivity. Whatever is the percentage of gap between the scanner that is the positivity of that particular Gemstone.

If one Gemstone is having positivity of less than 50% leave it and get a better stone.

Gemstone Compatibility

How to check through aura Scanner which Gemstone will suit you?

CHAPTER TWENTY-THREE

MANIK

Planet – Sun
Metal – Gold & Copper
Day – Sunday
Paksha – Shukla & Krishna Paksha
Finger – Ring Finger (Anamika)

If we check the Horoscope, if Sun is posited in the 6th or 8th, or 12th House or in Debilitated condition then wearing a Ruby will not be fruitful. The exception is Vipreet Raj Yoga. Ruby is generally fruitful for Leo, Aries, Saggitarius, and Pisces Lagna. But **it is not necessary that you need to know how to read the Kundli**. Without checking the Kundli also whatever result the aura scanner will give, will be exactly as present in your Aura, but yes if you know how to read birth charts then you can tally both for verifying. You can easily check the compatibility of Ruby Gemstone through Aura Scanner by following these steps :

1. Put the Mass Cable of the Aura Scanner on the person's photo printout.

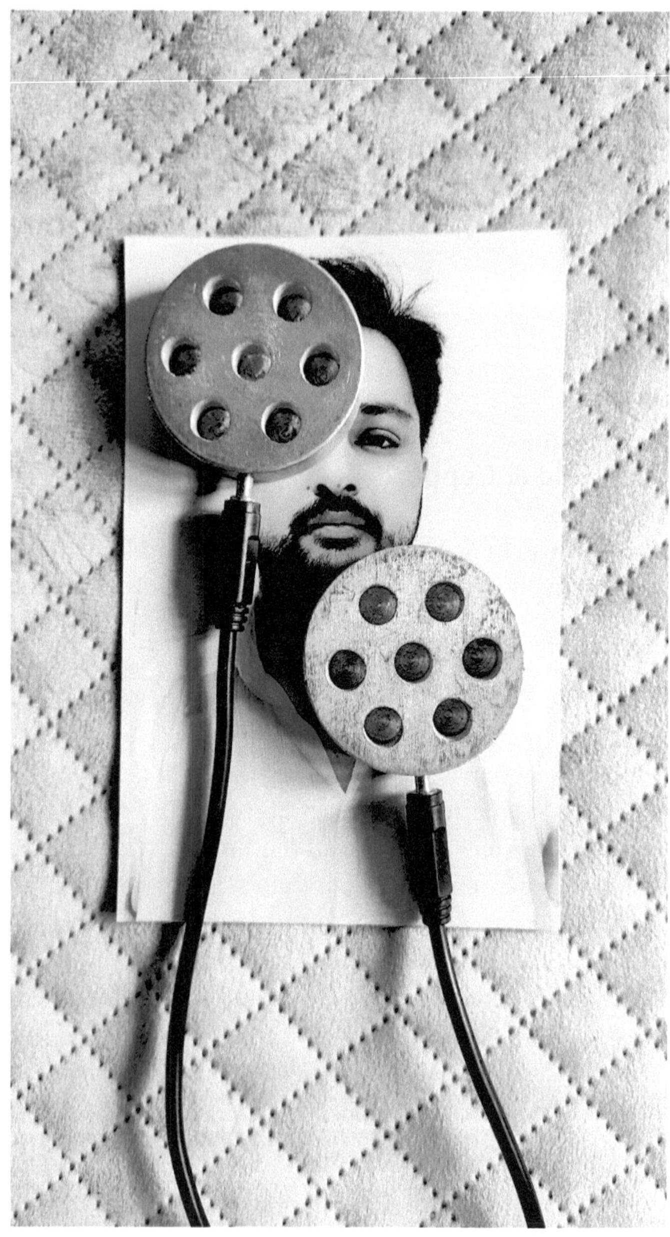

The mass cable on the Photo

2. Place the samples of Sun on both the limbs of the Aura Scanner.
3. Switch ON the Aura Scanner. Place both the limbs of the Aura Scanner at an 85° angle at the front.
4. Wait for the Reading and Note it down. (Suppose that the limbs make an angle of 50°)

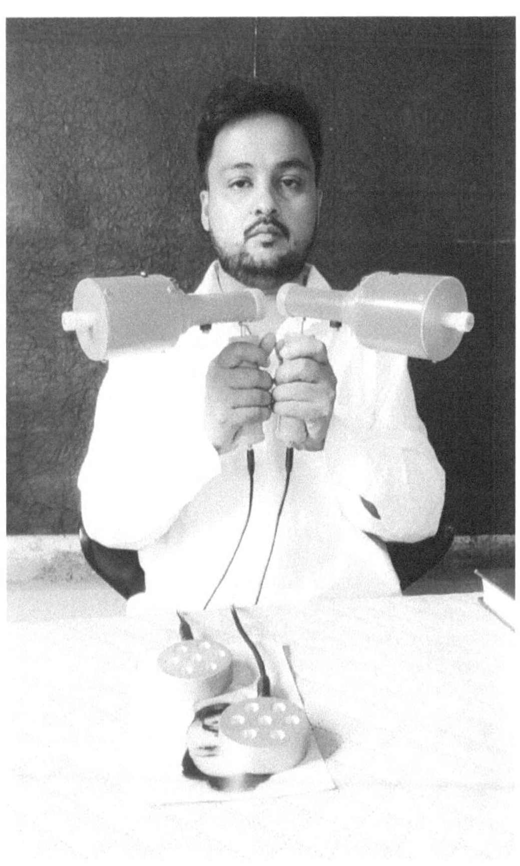

Scanning with the sample of SUN

5. Take one of the samples of Ruby and put it on the photo printout. You can even use an actual Ruby in place of the Ruby sample.

The sample of MANIK placed on the Photo

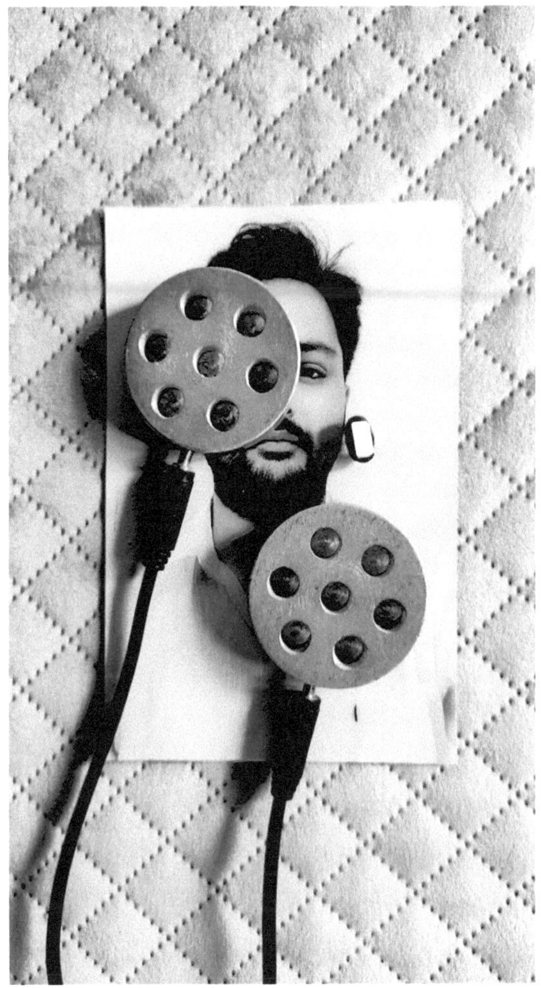

Manik stone placed on the Photo

6. Take the reading of the Aura Scanner, (Suppose that the limbs make an angle of 100°)

Aura Scanner at 100%

Conclusion – If the second reading is more, Ruby suits you. On the other hand, if the second reading is less than the first reading, Ruby will not suit you and you will only get bad luck after wearing it. Astrological purpose Ruby stone should never be worn in Silver.

CHAPTER TWENTY-FOUR

MOTI

Planet – Moon
Metal – Silver
Day – Monday
Paksha – Shukla Paksha
Time – A quarter of an hour after sunset
Finger – Smallest Finger (Kanishka)

If we check the Horoscope, if Moon is posited in the 6th or 8th, or 12th House or in Debilitated condition then wearing a Moti will not be fruitful. Moti is generally fruitful for Aries, Cancer, and Pisces Lagna. But it is not necessary that you need to know how to read the Kundli. You can easily check the compatibility of Moti Gemstone through Aura Scanner by following these steps(Refer to Chapter 21 for images):

1. Put the Mass Cable of the Aura Scanner on the person's photo printout.
2. Place both the samples of Moon on both the limbs of the Aura Scanner.
3. Switch ON the Aura Scanner. Place both the limbs of the Aura Scanner at an 85° angle at the front.

4. Wait for the Reading and Note it down. (Suppose that the limbs make an angle of 60°)
5. Take one of the samples of Moti and put it on the photo printout. You can even use an actual Pearl in place of the Moti sample.
6. Take the reading of the Aura Scanner, (Suppose that the limbs make an angle of 90°)

Conclusion – If the second reading is more, Pearl/Moti suits you. On the other hand, if the second reading is less than the first reading, Moti will not suit you and you will only get bad luck after wearing it. Astrological Pearls should never be worn in Gold.

CHAPTER TWENTY-FIVE

MUNGA

Planet – Mars
Metal – Gold & Copper
Day – Tuesday
Paksha – Shukla Paksha
Finger – Ring Finger (Anamika)

If we check the Horoscope, if Mars is posited in the 6th or 8th, or 12th House or in Debilitated condition then wearing a Moonga will not be fruitful. The exception is Vipreet Raj Yoga. Ruby is generally fruitful for Aries, Leo, Saggitarius, and Pisces Lagna. But it is not necessary that you need to know how to read the Kundli. You can easily check the compatibility of Moonga Gemstone through Aura Scanner by following these steps(Refer to Chapter 21 for images):

1. Put the Mass Cable of the Aura Scanner on the person's photo printout.
2. Place both samples of Mars on both limbs of the Aura Scanner.
3. Switch ON the Aura Scanner. Place both the limbs of the Aura Scanner at an 85° angle at the front.
4. Wait for the Reading and Note it down. (Suppose that the limbs make an angle of 60°)

5. Take one of the samples of Moonga and put it on the photo printout. You can even use an actual Red Coral/Moonga in place of the Moonga sample.
6. Take the reading of the Aura Scanner, (Suppose that the limbs make an angle of 90°)

Conclusion – If the second reading is more, Moonga suits you. On the other hand, if the second reading is less than the first reading, Moonga will not suit you and you will only get bad luck after wearing it. Moonga should never be worn in Silver(which is a common practice today). Fire element planet's stone should not be worn in a watery Metal, Silver. It will give you tension and health issues.

CHAPTER TWENTY-SIX

PANNA

Planet – Mercury
Metal – Silver & Gold
Day – Wednesday
Paksha – Shukla and Krishna Paksha
Finger – Smallest Finger (Kanishka)

If we check the Horoscope, if Mercury is posited in the 6th or 8th, or 12th House or Debilitated condition then wearing a Panna will not be fruitful. Mercury is generally fruitful for Gemini, Virgo, Capricorn, and Aquarius Lagna. But it is not necessary that you need to know how to read the Kundli. You can easily check the compatibility of Panna Gemstone through Aura Scanner by following these steps(Refer to Chapter 21 for images):

1. Put the Mass Cable of the Aura Scanner on the person's photo printout.
2. Place samples of Mercury on both limbs of the Aura Scanner.
3. Switch ON the Aura Scanner. Place both the limbs of the Aura Scanner at an 85° angle at the front.
4. Wait for the Reading and Note it down. (Suppose that the limbs make an angle of 60°)

5. Take one of the samples of Panna and put it on the photo printout. You can even use an actual Emerald/Panna in place of the Panna sample.
6. Take the reading of the Aura Scanner, (Suppose that the limbs make an angle of 90°)

Conclusion – If the second reading is more, Panna suits you. On the other hand, if the second reading is less than the first reading, Panna will not suit you and you will only get bad luck after wearing it.

CHAPTER TWENTY-SEVEN

PUKHRAJ

Planet – Jupiter
Metal – Gold & Brass
Day – Thursday
Paksha – Shukla Paksha
Finger – Index Finger (Tarjani)

If we check the Horoscope, if Jupiter is posited in the 6th or 8th, or 12th House or in Debilitated condition then wearing a Pukhraj will not be fruitful. Pukhraj is generally fruitful for Aries, Saggitarius, Leo, and Pisces Lagna. But it is not necessary that you need to know how to read the Kundli. You can easily check the compatibility of Pukhraj Gemstone through Aura Scanner by following these steps(Refer to Chapter 21 for images):

1. Put the Mass Cable of the Aura Scanner on the person's photo printout.
2. Place both samples of Jupiter on both the limbs of the Aura Scanner.
3. Switch ON the Aura Scanner. Place both the limbs of the Aura Scanner at an 85° angle at the front.
4. Wait for the Reading and Note it down. (Suppose that the limbs make an angle of 60°)

5. Take one of the samples of Pukhraj and put it on the photo printout. You can even use an actual Pukhraj stone in place of the Pukhraj sample.
6. Take the reading of the Aura Scanner, (Suppose that the limbs make an angle of 90°)

Conclusion – If the second reading is more than the first reading, then we can conclude that Pukhraj suits you. On the other hand, if the second reading is less than the first reading, Pukhraj will not suit you and you will only get bad luck after wearing it. In some cases, we can see there is hardly any difference in the readings, in that case, there is no need to invest so much for that planet's stone. Pukhraj should never be worn in Silver

CHAPTER TWENTY-EIGHT

DIAMOND

Planet – Venus
Metal – Silver or White Gold
Day – Friday
Paksha – Shukla and Krishna Paksha
Finger – Longest Finger (Madhyama)

If we check the Horoscope, if Venus is posited in the 6th or 8th, or 12th House or in Debilitated condition then wearing a Venus will not be fruitful. Venus is generally fruitful for Gemini, Virgo, Taurus, Capricorn, and Aquarius Lagna. But it is not necessary that you need to know how to read the Kundli. You can easily check the compatibility of Diamond Gemstone through Aura Scanner by following these steps(Refer to Chapter 21 for images):

1. Put the Mass Cable of the Aura Scanner on the person's photo printout.
2. Place samples of Venus on both limbs of the Aura Scanner.
3. Switch ON the Aura Scanner. Place both the limbs of the Aura Scanner at an 85° angle at the front.
4. Wait for the Reading and Note it down. (Suppose that the limbs make an angle of 60°)

5. Take one of the samples of Heera and put it on the photo printout. You can even use an actual Diamond gemstone in place of the Heera sample.
6. Take the reading of the Aura Scanner, (Suppose that the limbs make an angle of 90°)

Conclusion – If the second reading is more, Diamond suits you. On the other hand, if the second reading is less than the first reading, Diamond will not suit you.

CHAPTER TWENTY-NINE

NEELAM

Planet – Saturn
Metal – Ashtdhatu
Day – Saturday
Paksha – Shukla & Krishna Paksha
Finger – Longest Finger (Madhyama)

If we check the Horoscope, if Saturn is posited in the 6th or 8th, or 12th House or in Debilitated condition then wearing a Neelam will not be fruitful. The exception is Vipreet Raj Yoga. Neelam is generally fruitful for Taurus, Aquarius, Capricorn, and Virgo Lagna. But it is not necessary that you need to know how to read the Kundli. You can easily check the compatibility of Neelam Gemstone through Aura Scanner by following these steps :

1. Put the Mass Cable of the Aura Scanner on the person's photo printout.
2. Place both samples of Saturn on both the limbs of the Aura Scanner.
3. Switch ON the Aura Scanner.
4. Place both the limbs of the Aura Scanner at an 85° angle at the front.

5. Wait for the Reading and Note it down. (Suppose that the limbs make an angle of 60°)
6. Take one of the samples of Neelam and put it on the photo printout. You can even use an actual Neela/Blue Sapphire in place of the Neelam sample.
7. Take the reading of the Aura Scanner, (Suppose that the limbs make an angle of 90°)

Conclusion – If the second reading is more, Neelam suits you. On the other hand, if the second reading is less than the first reading, Neelam will not suit you and you will only get bad luck after wearing it. According to a few learned astrologers, Neelam gemstone should never be worn in Silver, Chandi. Rather wear it in Ashtdhatu than in Silver. Wearing Neelam in Chandi is the same as bringing sadhe-Sati upon yourself. Neelam is worn on any Saturday evening.

"After wearing Neelam either one becomes a King or a beggar", this statement is completely baseless in the same way as if a doctor suggests you take Calcium tablets that don't mean after the usage of Calcium tablets you will break the walls with one punch!

CHAPTER THIRTY

GOMED

Planet – Rahu
Metal – Silver
Day – Saturday
Paksha – Shukla & Krishna Paksha
Finger – Longest Finger (Madhyama)

Generally, Gomedh is a Gemstone that a common person should choose wisely. Even after checking the Kundli well known Astrologers also make mistakes when suggesting. You can easily check the compatibility of Gomedh Gemstone through Aura Scanner by following these steps(Refer to Chapter 21 for images):

1. Put the Mass Cable of the Aura Scanner on the person's photo printout.
2. Place both the samples of Rahu on both the limbs of the Aura Scanner.
3. Switch ON the Aura Scanner. Place both limbs together and at an angle of 85° from the front.
4. Wait for the Reading and Note it down. (Suppose that the limbs make an angle of 60°)
5. Take one of the samples of Gomedh and put it on the photo printout. You can even use an actual Gomedh/Hessonite in place

of the Gomedh sample.

6. Take the reading of the Aura Scanner, (Suppose that the limbs make an angle of 90°)

Conclusion – If the second reading is more, Gomedh suits you. On the other hand, if the second reading is less than the first reading, Gomedh will not suit you and you will only get bad luck after wearing it. Gomedh should never be worn for more than 6 months.

CHAPTER THIRTY-ONE

CAT'S EYE

Planet – Ketu
Metal – Silver
Day – Saturday
Paksha – Shukla & Krishna Paksha
Finger – Longest Finger (Madhyama)

In rare cases wearing the gemstone of Ketu, Lehsunia is recommended, but it is a tough thing to advise someone to wear Lehsunia. Still, you can easily check the compatibility of Ruby Gemstone through Aura Scanner by following these steps(Refer to Chapter 21 for images):

1. Put the Mass Cable of the Aura Scanner on the person's photo printout.
2. Place samples of Ketu on both limbs of the Aura Scanner.
3. Switch ON the Aura Scanner. Place both the limbs of the Aura Scanner at an 85° angle at the front.
4. Wait for the Reading and Note it down. (Suppose that the limbs make an angle of 60°)
5. Take one of the samples of Lehsunia and put it on the photo printout. You can even use an actual Cat's Eye/Lehsunia Stone in place of the Lehsunia sample.

6. Take the reading of the Aura Scanner, (Suppose that the limbs make an angle of 90°)

Conclusion – If the second reading is more, Lehsunia suits you. On the other hand, if the second reading is less than the first reading, Lehsunia will not suit you and you will only get bad luck after wearing it. Lehsunia should never be worn for more than 6 months.

CHAPTER THIRTY-TWO

DO SUBSTITUTE GEMSTONES WORK?

I will take the example of the Planet Venus. The Gemstone of Venus is a Diamond. As we know, everyone can not afford a Diamond so he is suggested to wear an Opal or to wear a White Sapphire. Suppose that when you check the compatibility of Opal with Aura Scanner you find that it will increase his Venus power by 20-30%. So he should wear an Opal stone, otherwise not. In the same way, other substitutes are checked. My observations are that in most cases substitute stones are favorable but Navratans will show much more effectiveness.

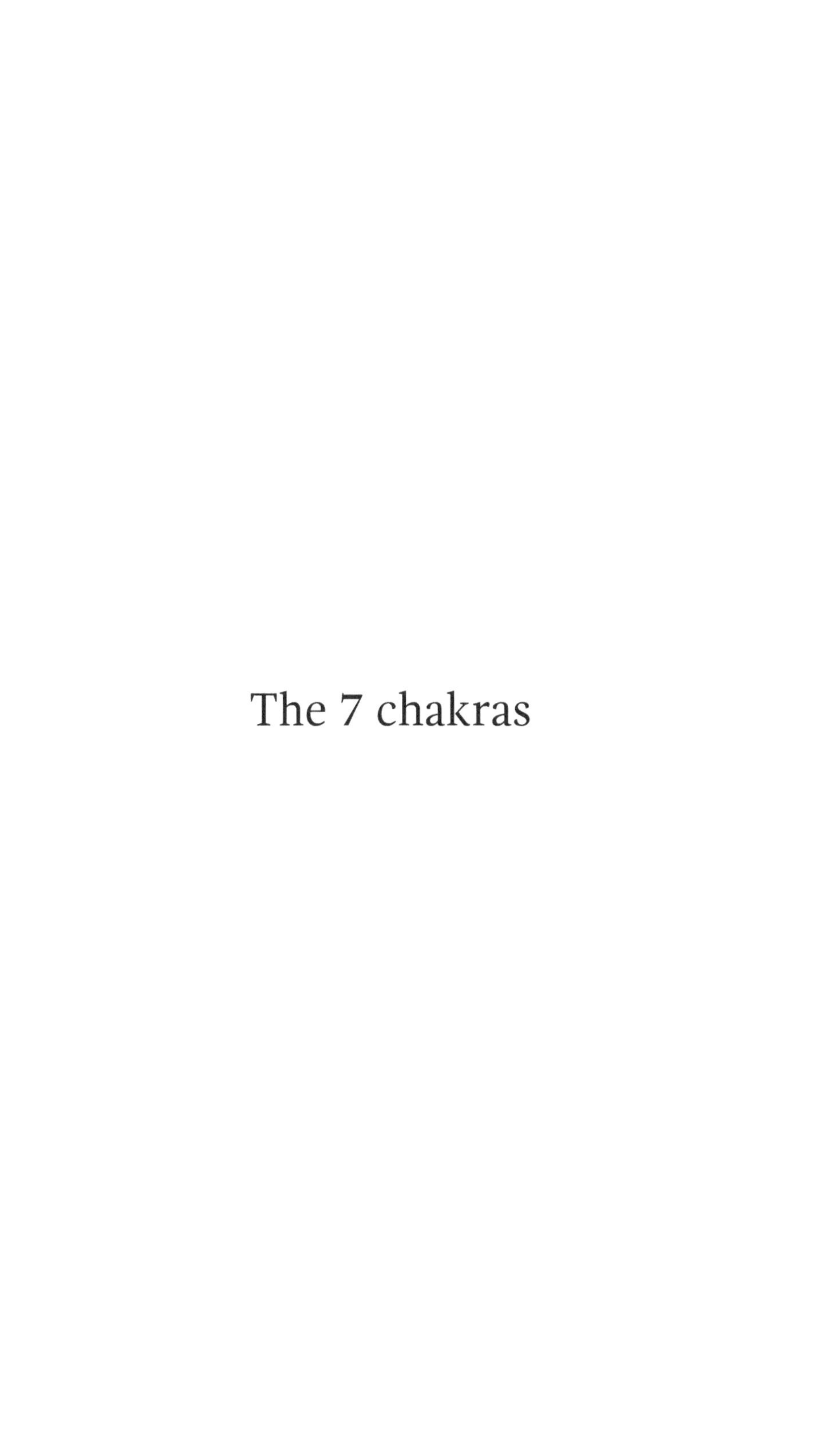

The 7 chakras

CHAPTER THIRTY-THREE

How to Diagnose Diseases

For checking the disease from aura you have to follow the proper steps to get the correct results. The first step should be to check all the Chakras one by one. The more the scanner opens for a Chakra the more problem is there in that region. For this, the calculation is just the opposite of the other samples. Suppose, the Mooladhar Chakra opens more than 40% then his Mooladhar Chakra is affected and he will face pain in the leg region. Also, he will face financial instability. In the Heart Chakra sample if the scanner opens more than 40% then that person is having many problems related to the heart region. Maybe his lungs are affected or he might be having Cholesterol and so on, he must surely get checked these things checked with a doctor. After this, we have separate samples for each disease. If a person has got Kidney stone the scanner will open in that situation. In most cases, the person who wants to check will be knowing it so then and there it will be checked for him.

CHAPTER THIRTY-FOUR

How to Check the Energy of Chakras

1. Mooladhar Chakra

The area included - the lower region of the body. Diseases - Anorexia, Knee pain, Back Pain, Piles, Blood Disease, Kidney Stone, Migraine, Obesity, and a few others. One or more of these diseases may affect the person if this chakra is imbalanced.

2. Swadhisthan Chakra

The area included - the intestine region of the body. Diseases - Drinking habit, Catching a cold often, Problems related to Reproduction, Petulance, Periods problem in females, Prostrate problem, Abortion. One or more of these diseases may affect the person if this chakra is imbalanced.

3. Manipura Chakra

The area included - the kidney, the Pancreas, and the Liver. Diseases - Anorexia, food allergy, Peptic Ulcer, Liver Problems, Diabetes, Smoking habits, Ulcer. One or more of these diseases may affect the person if this chakra is imbalanced.

4. Heart Chakra

The area included - the chest region of the body. Diseases - Asthma, Cough, Retardation, Heart Problem, High BP, Pneumonia, Sleep Disorder, Smoking. One or more of these diseases may affect the person if this chakra is imbalanced.

5. Throat Chakra

The area included - the throat region of the body. Diseases - Asthma, Bronchitis, Cold & Cough, Mental doubts, Black Cough, Tonsil, Pain in upper region of arms. One or more of these diseases may affect the person if this chakra is imbalanced.

6. Third Eye Chakra

The area included - The Eyes and the Ear region. Diseases - Habit of Forgetfulness, Sleep Disorders, Problem in the left eye, Tension, Sinus, Far-sightedness, Brain Tumor, Migraine. One or more of these diseases may affect the person if this chakra is imbalanced.

7. Crown Chakra

The area included - The Brain region. Diseases - Multiple Personality Syndrome, Habit of Forgetfulness, Fear, Migraine, Problem in Learning, Rtardness, Neurosis. One or more of these diseases may affect the person if this chakra is imbalanced.

CHAPTER THIRTY-FIVE

REMEDIES FOR CHAKRA IMBALANCE

If any Chakra is imbalanced then the use respective oil provided by us is most beneficial. I myself have used them so I can suggest others. My 3 chakras were imbalanced and it was visible in the readings also. But today, after the proper use after 2 months, the problem is less. Anyone who performs Aura Scanning can easily find it out.

There is separate oil for each Chakra which we provide in roll-on form. The method of usage is to apply it on the hands and put it below the cheek area. Then take the smell of the oil. this oil functions through smell organs. This process is to be done twice daily, continuously for at least 2-3 months. If the problem is in more than one Chakras then use the oils alternative day.

Note - On the same day, more than one oil should never be used.

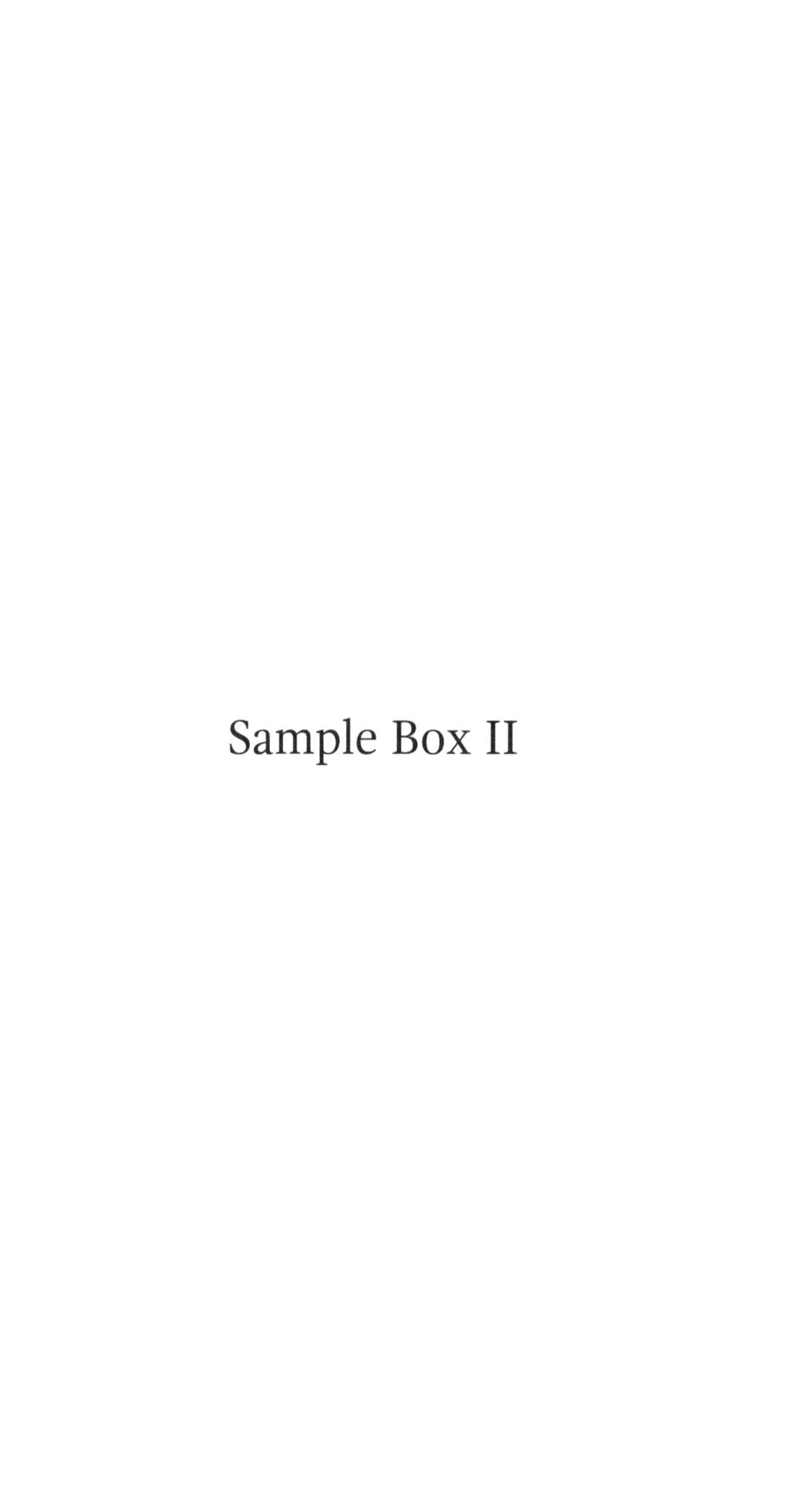

Sample Box II

Rudraksha

CHAPTER THIRTY-SIX

My first Aura Scan of Rudraksha

In my initial days of aura scanning, I tried to explore Rudraksha. I had two Rudraksh rosaries in my house. One of the rosaries was worn by my father and the other one was won by my elder brother. Both of them are still in excellent condition. I just wanted to know whether Rudraksha Rosary has got any power in it. Here I would like to mention that every Rudraksha Rosary is not suitable for everyone. The Rudraksha Rosary which suits me might not be fruitful for you. Obviously, first, test it and then wear it.

An interesting incident told by my father is that whenever by mistake he put on my elder brother's Rudraksh rosary he felt uncomfortable. Strange sensations arose in him that he is wearing something that do not belong to him. I thought I should definitely test the compatibility of both the rosaries with my dad's picture.

I took the mass-cable and put the first Rudraksh rosary on it which belonged to my elder brother. I wanted to check its positivity. I used the sample of Positivity and turned On the aura scanner. It gave me a reading of 90% which concludes that the rosary of my elder brother is having a positivity of 90%. Then I replaced Rudraksha Rosary with my father's rosary. The sample of positivity that was yet on the aura scanner. It gave the reading of a hundred percent. I was totally satisfied with the results. Now, the next and the most important part of the experiment came.

I took a postcard-size photograph of my father and then put the mass cable on the photograph. It was kept in such a way that the mass cable covered a little area of the photograph but was totally inside the photograph and no portion of those masses were outside it. Then I took the reading of positivity - 60%.

After that, I put my elder brother's Rosary the positivity of the aura scanner showed 40%. This shows that if my father will wear my elder brother's Rosary, it will decrease the Positivity of my father's Aura.

After this, I replaced my elder brother's rosary with my father's Rudraksha Rosary. This time the reading showed 90% Positivity. The conclusion is that my father's Rudraksh rosary is compatible with his own rosary but not with my brother's rosary. If anyone is curious to watch that video, I can provide the same. Any Rudraksha Rosary will suit you or not, you can check in the following steps :

1. Place the mass cable on the postcard-size photo printout.
2. Insert the sample of Positive in aura scanner.
3. Turn On the aura scanner and raise the limbs of the Aura Scanner at 85° angle at the front. Note the distance between both limbs. Suppose that the reading is 50%
4. Replace the Rudraksh rosary with the other rosary and take the second reading keeping the sample of positivity still in the scanner.
5. Note the second reading and turn Off the aura scanner.

Conclusion - If the second reading is more than the first reading that means that the Rudraksha Rosary is suitable for the person and he should definitely wear it. On the other hand, if the second reading is less than the first reading that means that using it will only decrease his positivity, so he should not wear it.

Place the Rudraksha in this way to check the Positivity of Rudraksha and its Mukha

CHAPTER THIRTY-SEVEN

WHICH MUKHI WILL SUIT YOU?

Which Mukhi Rudraksha Rosary will suit you is one of the toughest questions in astrology. In reality, 50% of astrologers are unaware of it. It is a general belief that 5 Mukhi Rudraksha better known as Panchmukhi Rudraksha suits almost everyone. But what about the other Mukhi Rudrakshas? We all know that one Mukhi Rudraksha is very very rare but do you know that it will not be beneficial for everyone? So how to check the compatibility of Rudraksha with your aura?

I have taken the example of 5 Mukhi Rudraksha the same procedure is to be followed with 1 to 14 Mukhi Rudrakshas.

Steps to be followed to check which Mukhi Rudraksha is best compatible for a person :

1. Put the mass cable on the photo printout of the person for whom the Rudraksh Mukhi compatibility is to be checked.
2. Let's take the example of 5 Mukhi Rudraksha. We will take the sample of 5 Mukhi. Insert the sample in both the aura scanner.
3. Turn on the aura scanner and wait for the reading.
4. Suppose the gap between both Limbs of the Aura Scanner is 20%.

Conclusion - The Rudraksha is not suitable for him and it will be of no use to him. In the same way, if the reading would be 40% or more than 40%, we could assume that Panchmukhi Rudraksha is suitable for him.

CHAPTER THIRTY-EIGHT

HOW TO CHECK IF ITS MUKHI IS CORRECT?

We all know the importance of the Mukha of Rudraksha. If a Rudraksha has 4 lines or 4 'Mukhas' it is a 4 Mukhi Rudraksha. If a Rudraksha has 7 lines or Mukhas then it is a 7 Mukhi Rudraksha and so on. This sounds quite simple but here comes the reality or the unfortunate reality of the sellers. Those sellers have expertise in forging an additional line to the Rudraksha beads. Let's say, we want to buy a 13 Mukhi Rudraksha, what the seller does sometimes is that he will try to convert a 12 Mukhi Rudraksha into a 13 Mukhi Rudraksha because he will get a higher price for a 13 Mukhi Rudraksha than a 12 Mukhi Rudraksha. For this, there is an interesting method to check the Mukhas of Rudraksha. Here I am taking an example of how to test whether 13 Mukhi Rudraksha is really 13 Mukhi or not. Follow the following steps :

1. Put the mass cable on one over the other, as shown in the picture. Then place that Rudraksha, which you want to test, over the mass cables.
2. Turn ON the Aura Scanner and join the scanners together and raise them at 85° at the front.

3. Note the reading from the angle of separation of both the limbs of the Aura Scanner. In this case, either the Aura Scanner will show a 180° angle or will not open at all. Observe the reading and just turn Off the Aura Scanner to conclude the reading.

Conclusion - If the Aura Scanner has opened to 180° that means the answer is Yes, it is 13 Mukhi. On the contrary, if the Aura Scanner does not open at all, then that means it is not a 13 Mukhi Rudraksha. In the same way, all 1 to 14 Mukhi Rudrakshas can be checked for confirmation. The Mukhi can also be tested by other sources like sending it to a lab for an X-Ray test. 13 Mukhi Rudraksha will have 13 seeds inside it which can be found on the X-Ray report easily. Nepali Rudraksha is found to be the most Energetic so if someone is selling you a Rudraksha that he says is a natural Nepali Rudraksha just check its positivity. If the positivity of the Rudraksha is less than 70% it cannot be a Nepali Rudraksha. This I have checked multiple times.

CHAPTER THIRTY-NINE

Benefits of Wearing Rudraksha

1 Mukhi Rudraksha -

The 1 Mukhi Rudraksha bead is believed to be a form of Lord Shiva, and it boosts the wisdom towards good faith. It also makes the person highly conscious. So if you want to fulfill the desires of your heart, you can start wearing a 1 Mukhi Rudraksha mala. This Rudraksha is also known for curi
ng migraines and other ailments in the body.

2 Mukhi Rudraksha -

2 Mukhi Rudraksha is a two-faced seed known as the 'Ardhanareeshwara' form of Lord Shiva. Wearing a mala made of this Rudraksha bead can bring peace and understanding, for example, between a husband and wife. It can also enhance the bond relationship between a brother-sister, mother-daughter, etc. If your mental focus is weak, you can start wearing this bead.

3 Mukhi Rudraksha -

3 Mukhi Rudraksha represents the fire God that is known as "Lord Agni" in Hinduism. It is believed that wearing the 3 Mukhi Rudraksha can erase the past karma or wash away the sins you've committed in your previous life. So, if you need to burn your 'bad karma,' get a 3 Mukhi Rudraksha mala.

4 Mukhi Rudraksha -

This seed is said to be associated with a similar deity from the Hindu lord known as Lord Brahma. He is also the supreme being in the Trimurti, which is the triple deity of supreme divinity. Lord Brahma, sometimes referred to as Brahman, is also known as the God of knowledge. According to the Mythology, he was forced toward this feat by his own wife, who grew frustrated that she was unable to find any inspired poets whose works she would like so much that she would be moved to tears. So, the powers of the 4 Mukhi Rudraksha are improving the wearer's memoto improventration, and overall intelligence.

5 Mukhi Rudraksha -

The 5 Mukhi Rudraksha represents the five elements – the sky, air, fire, water, and earth. The ruling planet of this Rudraksha is Jupiter, and it is the most popular Rudraksha bead among other beads. It brings a lot of benefits like curing a plethora of health issues that include acidity, flatulence, blood pressure problems, etc. In addition, it boosts the wearer's confidence and makes the mind more powerful.

6 Mukhi Rudraksha -

Lord Kartikeya was the son of Lord Shiva and is the deity of 6 Mukhi Rudraksha. It is believed that the wearer of the 6 Mukhi

Rudraksha can keep jealousy, anger, and mental excitement under control. It can also bring pleasure, comfort, and happiness to one's life. Apart from that, the 6 Mukhi Rudraksha can also help treat sexual problems, urinary problems, kidney failures, indigestion, and a lot more.

7 Mukhi Rudraksha -

7 Mukhi Rudraksha is highly auspicious, with Goddess Mahalakshmi as its deity and Saturn being its ruling planet. It symbolizes the seven sacred rivers and is believed to cure the wearer of any kind of poison. If you've been dealing with bad luck for a long time but can't find a cure, it is suggested that you must wear a 7 Mukhi Rudraksha mala. Wearing it can also bring good luck in court cases and successful business ventures.

8 Mukhi Rudraksha -

If you cannot focus on achieving your goals due to never-ending obstacles, wearing an 8 Mukhi Rudraksha mala can get rid of them for you. Other than that, it boosts the intelligence and analytical thinking of the wearer. In addition, the wearer gets the name, fame, leadership qualities, self-control, and a lot more benefits. Coming to the health benefits, wearing an 8 Mukhi Rudraksha mala can also prevent a plethora of diseases related to the nervous system, gallbladder, and prostate gland.

9 Mukhi Rudraksha -

With the ruling deity Goddess Durga and the ruling planet Ketu, 9 Mukhi Rudraksha gives you the benefit of pleasing all nine planets of the horoscope. As a result, the wearer lives a stress-free life and escapes untimely death. It also bestows the wearer confidence and courage enough to deal with any obstacle in life.

10 Mukhi Rudraksha -

Ten mukhi Rudraksha is helpful in curing sexual disorders. 10 mukhi Rudraksha wearer cures from anxiety, worries, Back pain and psychic attacks. Das mukhi Rudraksha blesses the wearer with peace and protection in all aspects of life

11 Mukhi Rudraksha -

It is believed that the person wearing an 11 Mukhi Rudraksha will live an adventurous life. There's no way he or she dies an untimely death and tastes the fruits of success for sure. But, apart from that, it enhances the decision-making abilities of the wearer, removes the problems of yogic practices, and gets rid of various health problems too.

12 Mukhi Rudraksha -

12 Mukhi Rudraksha beads help you to connect with the root chakra, which is the area where most of our problems arise. 12 Mukhi Rudraksha is prevalent in nature and can help you remain grounded throughout your life. You would have a healthy life, with no money problems. It is also believed that the wearer is protected from all diseases and all forms of evil. Wearing a 12 Mukhi Rudraksha will enhance your relationships with others, give your children a successful career, and add positivity to every aspect of your life.

13 Mukhi Rudraksha -

The Terah Mukhi Rudraksha helps the wearer in solving problems from sexual diseases, Low libido, low sperm count/infertility etc. 13 mukhi Rudraksha wearer heals chronic low back pain, sciatica, pelvic pain, kidney related problems and muscle cramps.

14 Mukhi Rudraksha -

The biggest stress comes from the mind and the soul in the body. A 14 Mukhi Rudraksha at the center of the forehead is supposed to cool the brain, making a person more relaxed, creative, and positive. It helps in overcoming anger, jealousy, and other self-destructive traits. In the soul, it helps a person face challenges and come out victorious. At the forehead, the Rudraksha is supposed to help fight depression, stress, and migraines.

Aura Scanner & Vastu

CHAPTER FORTY

REMOVE VASTU DOSHA

Vastu Dosha Remedies

Vastu Dosha is of two types. Either it is a constructional Vastu Dosha caused due to wrong construction of rooms or it is due to the wrong position of objects in the house. Taking the example of constructional Vastu Dosha if there is an extension plot or a portion of the plot is cut it will cause Vastu dosha. The remedy constructional Vastu dosh is simple and effective – use the multipurpose rod provided by Mangalam Vastu. In the same way, if the plot is extended then also in a few directions it will cause Vastu Dosha which can be neutralized by the use of this multi-purpose rod. It is to be noted that the North East extension of the plot is very fruitful, North Extension is also quite favorable and after that comes the South East extension but that is not so auspicious, and so on. It can be checked easily with the help of the Aura Scanner.

The cut plot in North East direction is unfavorable and also the cut plot in the North direction is not good. The family should leave that plot as soon as possible but if not able to leave then there is a simpler remedy – use a multipurpose rod manufactured by Mangalam Vastu, which is capable of reducing the side effects of Vastu Doshas. Talking about efficiency, one can compare the positivity of these rods with other rods available in the market, and easily one can find the difference. If I recommend someone a product it is because I want to help the reader so getting rods from many poor-quality manufacturers will not solve the purpose.

How to find out Vastu Doshas in the house

Suppose you have Aura Scanner and a client calls you to the constructed house to know which place is best for which purpose. You can easily check and guide him correctly. Even you can suggest to him also the color paint that will be best for which wall and room. There is no scope of negativity to get in the house if one uses this Aura Scanning machine properly. The family members will find themselves happy from the inside and support each other for a better life.

How to check which color paint is best for a wall

1. Take a sample of the color that you want to paint on the wall. Insert both samples into the limbs of the Aura Scanner in the front.

1. Go and stand facing the wall at a distance of roughly 2 feet. Then turn ON the Scanner and lift the Aura Scanner at an angle of 85° as shown in the picture.

Aura Scanner will show you in percent which color is suitable for that wall and which is not at all suitable. If the Scanner opens to at least 60% percent for a color then you can use that color paint on the wall. In the same way, if the Aura Scanner shows a reading less than 30% then reject that color's usage for that particular wall and check that color with other walls. The yellow color in the West direction wall is good. In the South-East Red color is good & favorable but the same Red color will be highly unfavorable in the North wall. Instead, a light Blue color or White color both are good for the North side wall of a house. This may vary from home to home so my advice is that get the help of the Aura Scanner instead of just doing a Google Search. Google will tell you a standard answer but Aura Scanner will give you the answer which is best suitable for your house specifically.

The procedure of finding out Vastu Dosh in Home

1. Find out the Centre of the building by putting the Sample of the Centre in both the limbs of the Aura Scanner. You must be knowing which room falls to the center of the construction, just go to that room and operate the Aura Scanner to find the center of that middle room. This can be an easy process once you get used to operating the scanner. The Scanner will direct you to go Right or Left or Front and when you reach the Centre point of the room it will open to 100%. That place is the construction Centre of the Home also known as the Brahm Sthaan.

2. Insert the sample of North from Sample box – 2 and lift the Aura Scanner at an angle of 85° at the front. The scanner will point towards the Geographical North of itself. I mean to say that if you are facing towards the East and you put on the sample of East then this scanner will point towards North! Any mobile Compass app will also show you the North direction so you can get a confirmation. So for checking the Vastu Dosha of the North direction first you have to face towards perfect North at the Brahm Sthaan of that house.

3. Again take a reading from the sample of North and this time you have to give the command to the scanner to show the percentage of Vastu Dosha in the North direction.

   Speak aloud – “North me Kitna percent Vastu Dosh hai?”

Suppose the scanner opens to 60% it means that the North direction of that house is affected by Vastu Dosha to 60%.

Immediate Vastu Dosh correction for the North is required.

4. Standing at the Brahm Sthaan note that all things are placed in the North direction. There might be a Bed, a T.V and the wall color might be Green. Now separately you need to check all those 3 things by using the respective samples. There are samples of T.V and Bed so it can easily be checked. Then for a color check, if you don't have the actual color of the wall you can check with the sample of Positivity how much is that color suitable for that wall. You just have to speak out the command for the Aura Scanner, "How much percentage of this color is good for this wall?"

5. Whichever object is not good in result note that on a paper. After that from the center of that room gives a command – "Which is the best direction for a Bed in this room". The Aura Scanner will lead you to that corner of the room that is best for the Bed.

In this way when you move the faulty object in the right direction then the Vastu Dosha will also disappear which can be confirmed at last by giving the command once more.

What we did for the North, is the same process we have to do for the East, West, and South also. At my home when I checked for the North direction Vastu Dosha it was more than 50%. Then I found that the Bed in the North was not posited well, also the Almirah was not in the correct direction. When I checked for the mirror standing at the center of the room then it showed Vastu Dosha of 40%. So I gave a command to the scanner, "Which direction is best for Mirror in this room?" The Aura Scanner pointed toward the North. I turned North and walked to the wall in the North direction. There it opened to 100%. I thought I should Google Search for confirmation because I didn't know in which direction we should place a mirror in our room.

I was shocked because Google also says that the perfect direction for Mirror placement is North and East. How accurate is this Aura Scanning machine..! and then there is Wikipedia, which sadly, still disbelieves in Aura science! Actually in our house mirror placement was originally at the same place which Aura Scanner was showing so the family members also said “Sahi hai, Sheeshe ka Asli jagah to wahi tha. Ab dubara wahin aa gaya”.

CHAPTER FORTY-ONE

HOME VASTU

How to remove Constructional Vastu Dosh at Home

The constructional Vastu Dosha has two solutions either you can use Directional rods provided by us or you will have to demolish and re-construct some parts of the house. Suppose that the North-East is used for urinal purposes then there is no remedy for it. You will have to lock up the Toilet. North - East Toilet is never giving good effects and causes health issues, people have to visit hospitals more often but yes, you can still use that area for washing clothes and bathing. Just toilet activity can't be done at that place because it is the Head portion of the Vastu-Purush and if the head is injured, there is no remedy. So, North-East is most fruitful if used as a Worship Room. You will find peace, prosperity and good Luck in the house. Except for North-East, there is a remedy for every portion of the house. The steps to detect Vasu Dosha are given in the next Chapter.

Once Vastu Dosha is detected, suppose there is Vastu Dosha in the Kitchen and it is constructed in the wrong direction such as North, you can go near the Gas Stove and check with the sample of the Gas Stove. If it opens less than 50% then there is Vastu Dosha and needs remedy. Here what you can do is either balance the South or place a Multipurpose rod provided by us. This rod is to be placed there below the Gas Stove and can either be kept on the ground or if the construction is still going on it can be buried under the ground also. Both ways it will receive either the Air Element or the Earth Element and will work well.

The next step is the reconfirmation of the Vastu Remedy we have made. Take the reading of the Gas Stove once again by using the Gas Stove sample, after you have placed the multi-purpose rod. The aura scanner will surely open up to more than 85% - 100% indicating that the remedy was successful.

My personal experience regarding the Vastu Dosh remedy

Vastu Dosh can easily be detected and at least positional Vastu Remedy can be done by anyone. One may take the help of the Aura Scanning machine, one can call an expert for help or one can take the task of self-help and get help from the internet. Suppose one wants to recheck his observations and conclude the scanning then one has to take the help of the traditional method for which one has to obtain a proper Map of the Home which we call in Hindi, Naksha. Naksha or plan is made by the Engineer or Vastu Consultant by paying an amount of about two thousand rupees and it will be accurate. One has to take the help of a local Engineer in the area and get the proper plan made of the house and then get it printed for analysis.

Get a Shakti Chakra from Amazon and place it at the exact degree as on the Entrance. I mean to say that go and stan at the Entrance of the House and open a compass and note down what degree is it? The Map might be showing 280°. So this 280° you have to place the Shakti Chakra on the Map and whatever result you get there will be the most accurate. I will explain this with my personal experience but unfortunately, I am not able to show openly my own house‘ map due to privacy issues I'm revealing the process here.

Diagonals on the map have to be made from all corners of the Map.

Mark the Centre of the Map. The place where the Diagonals meet will be the Centre point and it is known as the Brahm Sthaan. Brahm Sthaan should be left unoccupied. Try to keep it as light s possible for the property. The Brahm Sthaan can be used as Hall or open area and never for Kitchen, Toilet, Dining etc. Place the Shakti Ckara's Centre on this Brahm Sthaan and rotate it to the Degree we took at the Entrance. Now the picture will be very clear. You can check now for the Toilets, Bedrooms, Kitchen and Puja Room. Suppose Bed Room 1 falls to the W2 portion on the shakti Chakra you can check for that on the internet if it is fruitful or not. Suppose the Enterance falls on the E3 then you can check on the internet how is E3 Enterance good/Bad. E3 Entrance is most auspicious so it is fine. If Enterance is not good take the help of Aura Scanner and read the Energy of your Entrance!

How to remove Positional Vastu Dosh at Home

Bed Rooms

There are 3 types of Bed Rooms in a Home - Master Bed Room, Children Bed Room and Guest Bed Room. Master Bed Room is for the owner of the house. The children's bedroom is for the wellness and growth of school-going kids and the Guest bedroom is either for guests or the butler. The correct bedroom must be used by the correct members of the house. For eg., if a Guest bedroom is used by the owner of the house then he will be forced by destiny to go for tours very often and he will live in his own house the same as a Guest. In the same way, if a servant of the house is given the Master bedroom unknowingly then he will get the attitude to rule the house. Every person will be bound to follow the servant's opinion. You might tell him to cook Rice Daal but he will not listen and make some item of his wish so this causes tension in the family. With the help of this Aura Scanner, we can find out which bedroom should be used for which purpose.

1. Master Bed Room -

To find out the Master Bed Room there is a sample f Bed Room in Sample Box – 2 of the Aura Scanner. Just stand in front of the bedroom. Insert those samples in the Aura Scanner and raise it to 85°. Then give the command -

"For how much percentage is this Bed fit for Master Bed?"

the scanner will first move left and right showing that it is ready to analyse the Aura and then we have to give this command by speaking it out to ourselves. The scanners will go apart to a certain angle from each other. 180° gap between the scanners means it is 100% fit for the master bedroom. A 0° gap between the scanner means we cannot use this bed for the master bedroom and so on. If we get negative results in this room then perform the same steps in another bedroom. Whichever bedroom you get a maximum reading on the Master Bed-Room command you should use it as Master Bed Room.

2. Children Bed Room -

For children's bedrooms also the same process s repeated. In this, you have to use the sample of Bed Room and go to any Bed Room and give a command,

"How much percentage of this Bed Room is fit for Children's bedroom ?"

Whichever room gives the highest reading is to be used by children for sleeping.

3. Guest Bed Room -

The same process as told above is to be repeated but this time the command will be,

"How much percentage of this BedRoom is fit for Guest Bed Room?"

Among the 3 bedrooms, one of the bedrooms will surely give a very high reading on the command. So in this way, we can use the respective bedroom for our daily sleep.
The guest bedroom's ideal direction is North-West. A girl which is at the age of marriage should be given a North-West direction bed room because very soon you have to do her Kanya Daan and it will happen also but if unknowingly you make her sleep either in the West or in the North direction then her marriage will be delayed.

Kitchen

In the kitchen, there are important things like the Gas Stove, Wash basin, Microwave Owen and Refridgerator. One by one you can use all these samples from the Sample Box 2 to find out which place is best from which object. First, you can take readings for Gas Stove by using that sample in the Aura Scanner. You need to take this reading from the centre of the kitchen and wait for the scanner to show you the idle direction for Gas Stove. Then after that check for tap water, microwave etc. one by one. Also after that, you can check for Fire Element and Water element if the reading is high then they are balanced well otherwise you'll have to use a multipurpose Rod to neutralize the energies. This multipurpose rod just needs to b placed there and after then when you take the reading both elements will show 100% readings. Sometimes the Gas Stove and Tap Water are too close to each other. This creates Vastu Dosha so it needs to be checked and removed.

Pooja Room

One of the most important Vastu Dosha is the one obtained from Puja Room or worship area. There should be no damaged idol in the worship room and not only that but also the direction of the Puja Room constructed should be in idle way. Sometimes if you have a broken idol in the puja room then we can easily scan its energy from this aura Scanner. For that there is a process:

1. Inset the sample of MANDIR in the scanner and whether the direction of the Puja Room is correct.

2. Insert the sample of Positivity and go in front of every idol one by one. In most cases, the Energy will be up to 100% but if it is less than 50% in this case just take out that idol from there.

3. Give the command to the scanner which is the best direction of the idol. Either the scanner will point towards a particular direction or it will not show any reading. If the scanner doesn't show any direction then it means that the idol is broken from someplace and you can drown that idol in a river (Visarjan).

There is a way in which you can check all the objects in your house. Just use that sample from Sample Box – 2 for eg. For checking the Washing machine's Energy in your home just insert that sample in this Aura Scanner and take the reading. I then Aura Scanner shows opens up less than 60% then which means that you should choose a better location for that washing Machine. You can give a command -

"Which is the best direction for Washing Machine?"

The scanner will take you to that location of the house. The washing Machine has Mars, Rahu and Moon's effect in it because Mars denotes Machinery, Rahu denotes dirt which comes out from the Washing Machine and then there is the Moon's role because of Water. It is believed that if Washing Machine is not placed in the correct direction or worse if it is out of order and left idle then that can cause quarrels in the family. So one should surely check these things like T.V, Fridge, A.C, Computer, Washing Machine, Sofa set, dining table, Inverter etc. There is a list of all the items in the Vastu Sample Box No. 2 in the starting chapter of this book so you can refer to that for better understanding. If your house Children's Study Table is not posited well then kids might be not interested in studies. When you change the direction with help of Aura Scanner then you'll see visible changes in their education and attitude. Aura Scanner is a boon to this modern world.

CHAPTER FORTY-TWO

Industrial Vastu

Application

In industries, there will be Raw Materials, some goods will be manufactured, and also there will be a place for Goods storage. If we place the Ready goods at a particular place that is influenced by Air Element then as per Vastu those goods will be sold off in a very less period of time. In the same way, there are samples of Boss Sitting & Staff Sitting. If the Boss sitting is right then the Boss' command will be full of influence. On the other hand, if the staff is placed in the place that should have been used by the Boss for his cabin then it is found that the staff will be able to impress the Boss easily to work as per his intentions. Then we can check the lift for the Goods area, the Entrance of the Industry and so on. All these sitting arrangements should be in a proper place.

Aura Scan Report

CHAPTER FORTY-THREE

How to Generate an Aura Scan Report

There is a protocol/system which should be followed for correct analysis.

Some answers will be in Yes/No while other answers come in percentage.

The results of Chakras and Diseases should be analyzed in just the opposite way, i.e the more the reading comes the more problematic it is. Others are understood ordinarily, the more the gap between both the scanners, the more percentage it is and the more favorable it is.

Given below is the Aura Scan report that an astrologer should generate:

Positive - __%

Negative - __%

U.V - / Nazar Dosh – Yes / No

Bandhan Dosh – Yes / No

Pitra Dosh – Yes / No

Kalsarp Dosh – Yes / No

P.Y – Yes / No

Manglik Dosh – Yes / No

Marriage Yog – Yes / No Time -

Santan Yog – Yes / No Time -

Mooladhar Chakra - __%

Sacral Chakra - __%

Manipur Chakra - __%

Heart Chakra - __%

Throat Chakra - __%

Third Eye Chakra - __%

Crown Chakra - __%

Disease - ______

Sun - __%	Manik - __%
Moon - __%	Moti - __%
Mars - __%	Munga - __%
Mercury - __%	Panna - __%
Jupiter - __%	Pukhraj - __%
Venus - __%	Diamond - __%
Saturn - __%	Neelam - __%
Rahu - __%	Gomed - __%
Ketu - __%	Lehsunia - __%

Geopathic Stress – Yes / No

CHAPTER FORTY-FOUR

LIST OF SAMPLE BOX I

Planets & Positivity

1. Positive
2. Negative
3. U.V -
4. Bandhan Dosh
5. Pitra Dosh
6. Kalsarp Dosh
7. P.Y
8. Manglik Dosh
9. Marriage Yog
10. Santan Yog
11. Mooladhar Chakra
12. Sacral Chakra
13. Manipur Chakra
14. Heart Chakra
15. Throat Chakra
16. Third Eye Chakra
17. Crown Chakra
18. Low BP

19. High BP
20. Heart Problem
21. Sugar
22. Liver
23. Bone Disease
24. Migraine
25. Cervical
26. Lungs
27. Asthma
28. Kidney Problem
29. Kidney Stone
30. Gall Bladder Stone
31. Joint Pain
32. Nervous System
33. Thyroid
34. Cancer
35. Sinus
36. Acid Problem
37. Uric Acid
38. Cholesterol
39. Intestine
40. Blood Disease
41. Orange
42. White
43. Red
44. Green
45. Yellow
46. Off White
47. Blue
48. Black
49. Brown
50. Sliva
51. Manik
52. Moti
53. Munga

54. Panna
55. Pukhraj
56. Heera
57. Neelam
58. Gomed
59. Cat's Eye
60. GEO
61. Sun
62. Moon
63. Mars
64. Mercury
65. Jupiter
66. Venus
67. Saturn
68. Rahu
69. Ketu
70. Centre

Sample Box - 1

CHAPTER FORTY-FIVE

LIST OF SAMPLE BOX II

Vastu

1. 1 Mukhi
2. 2 Mukhi
3. 3 Mukhi
4. 4 Mukhi
5. 5 Mukhi
6. 6 Mukhi
7. 7 Mukhi
8. 8 Mukhi
9. 9 Mukhi
10. 10 Mukhi
11. 11 Mukhi
12. 12 Mukhi
13. 13 Mukhi
14. 14 Mukhi
15. Water element
16. Air Element
17. Fire Element
18. Space Element

19. Earth Element
20. Washing Machine
21. Important Paper
22. Invertor
23. Refridgerator
24. Mandir
25. Almirah
26. Main Gate
27. Locker
28. Bed
29. Dining Table Gerator
30. Generator
31. Bore Well
32. Sofa Set
33. Stair Case
34. Study Table
35. Gas Stove
36. Micro Wave
37. Water Tap
38. Toilet Seat
39. Mirror
40. Waste Material
41. Raw Material
42. Finished Goods
43. Machine
44. Lift For Goods
45. Staff Sitting
46. Boss Sitting
47. Computer
48. Lift
49. East
50. West
51. North
52. South
53. NE

54. NW
55. SE
56. SW
57. ENE
58. WSW
59. WNW
60. NNE
61. SSW
62. SSE
63. ESE
64. - 72. Numbers 1 - 9

Sample Box - 2

CHAPTER FORTY-SIX

LIST OF SAMPLE BOX III

45 Devtas, 32 Dwaar

Shikhi (NE2)
Parjanya (NE3)
Jayant (E1)
Mahendra / Indra (E2)
Surya (E3)
Satya (E4)
Bhrusha (E5)
Akash (SE1)
Anil (SE2)
Pushaa (SE3)
Vitatha (S1)
Gruhakshata (S2)
Yama (S3)
Gandharva (S4)
Bhringraj (S5)
Mrig (SW1)
Pitra (SW2)
Dauwarik (SW3)
Sugreeva (W1)

Pushpadnt (W2)
Varun (W3)
Asura (W4)
Shosha (W5)
Paapyakshama (NW1)
Rog (NW2)
Naag (NW3)
Mukhya (N1)
Bhallaat (N2)
Som (N3)
Bhujang (N4)
Aditi (N5)
Diti (NE1)

Sample Box - 3

Special Mentions:

Ambresh & Akshay Arya
have been fulfilling the demand for all kinds of Natural, unheated & untreated Gemstones, wearing which people's life has changed!

It is not just a matter of likeness, it is a matter of Trust.
Visit the Deoghar store now!

 BIS CERTIFIED HALLMARK SHOWROOM

8825236196 / 9572403724

Jalsar Road, Near Mahavir Akhada, B. Deoghar (Jh.)

www.ingramcontent.com/pod-product-compliance
Ingram Content Group UK Ltd.
Pitfield, Milton Keynes, MK11 3LW, UK
UKHW042018190726
13854UKWH00005B/2349